About Island Press

Since 1984, the nonprofit Island Press has been stimulating, shaping, and communicating the ideas that are essential for solving environmental problems worldwide. With more than 800 titles in print and some 40 new releases each year, we are the nation's leading publisher on environmental issues. We identify innovative thinkers and emerging trends in the environmental field. We work with world-renowned experts and authors to develop cross-disciplinary solutions to environmental challenges.

Island Press designs and implements coordinated book publication campaigns in order to communicate our critical messages in print, in person, and online using the latest technologies, programs, and the media. Our goal: to reach targeted audiences—scientists, policymakers, environmental advocates, the media, and concerned citizens—who can and will take action to protect the plants and animals that enrich our world, the ecosystems we need to survive, the water we drink, and the air we breathe.

Island Press gratefully acknowledges the support of its work by the Agua Fund, Inc., Annenberg Foundation, The Christensen Fund, The Nathan Cummings Foundation, The Geraldine R. Dodge Foundation, Doris Duke Charitable Foundation, The Educational Foundation of America, Betsy and Jesse Fink Foundation, The William and Flora Hewlett Foundation, The Kendeda Fund, The Andrew W. Mellon Foundation, The Curtis and Edith Munson Foundation, Oak Foundation, The Overbrook Foundation, the David and Lucile Packard Foundation, The Summit Fund of Washington, Trust for Architectural Easements, Wallace Global Fund, The Winslow Foundation, and other generous donors.

The opinions expressed in this book are those of the author(s) and do not necessarily reflect the views of our donors.

Local Climate Action Planning

Michael R. Boswell, Adrienne I. Greve, and Tammy L. Seale

Local Climate
Action Planning

Michael R. Boswell, Adrienne I. Greve, and Tammy L. Seale

Images by Dina Perkins

ISLANDPRESS

Washington | Covelo | London

© 2012 Michael R. Boswell, Adrienne I. Greve, and Tammy L. Seale (text),
Dina Perkins (images)

Library of Congress Cataloging-in-Publication Data

Boswell, Michael R.
  Local climate action planning / Michael R. Boswell, Adrienne I. Greve,
and Tammy L. Seale.
     p.   cm.
  ISBN-13: 978-1-59726-961-2 (cloth : alk. paper)
  ISBN-10: 1-59726-961-1 (cloth : alk. paper)
  ISBN-13: 978-1-59726-962-9 (pbk. : alk. paper)
  ISBN-10: 1-59726-962-X (pbk. : alk. paper) 1. Climate change mitigation—
Planning. 2. Climate change mitigation—Government policy. 3. Greenhouse gas
mitigation—Planning. 4. Greenhouse gas mitigation—Government policy.
5. Communication in the environmental sciences. 6. Communication in social action.
I. Greve, Adrienne I. II. Seale, Tammy L. III. Title.
  QC903.B67 2011
  363.738'746—dc23
                                    2011036255

Printed on recycled, acid-free paper

Manufactured in the United States of America
10 9 8 7 6 5 4 3 2 1

Keywords: California climate change legislation, carbon footprint, climate change
adaptation, climate change mitigation, community planning, comprehensive plan,
emission reduction, energy efficiency, global warming, greenhouse gas emissions,
natural hazard mitigation, public participation, resilient communities, smart growth,
transportation planning, visioning

For our nieces and nephews

Contents

Preface

In response to increasing evidence that climate change is occurring and has the potential to negatively impact human civilization, climate action plans are becoming the primary comprehensive policy mechanism for the reduction of greenhouse gas emissions and for management of risks posed by climate change (climate change *adaptation*). Climate action planning is an opportunity: an opportunity for communities to control their destinies in the face of global change, to achieve energy security, to sustainably develop their economies, and to ensure a high quality of life. Communities can seize this opportunity by building on existing planning and partnerships, by being creative and innovative, and by committing to working together for a better tomorrow for themselves and the next generations.

This book describes the process and methods for preparation of climate action plans for local governments. It is intended to be a practical guide; helping readers to navigate the principal actions and critical considerations for developing a climate action plan for their local jurisdiction. We believe that the best climate action plans are based on sound science, public education and outreach, recognition of global context and external constraints, awareness of the interdependent nature of local policy, and integration with existing planning policies and programs. We base this on our professional experience of working on over three dozen climate action plans and greenhouse gas emissions inventories in California and our academic experience researching and publishing on the state of climate action planning practice nationwide. As of this writing, there is no book that addresses climate action planning as a specific area of professional practice. Our hope is to contribute to the robust development of this professional field.

The book is aimed primarily at those who have been tasked with preparing a climate action plan, whether they are local government staff

xii   Preface

members, consultants, or community volunteers. Professionals who
should find the book useful include city/urban planners, regional planners, land use planners, environmental planners/managers, transportation
planners, city administrators, city attorneys, city engineers, emergency
managers, public works and transportation managers, architects, landscape
architects, building officials, sustainability coordinators/managers, and
climate action managers. In addition, the book is accessible to students
in these fields and can serve as an introductory text to the field of climate
action planning. The book should also be useful to anyone involved or
interested in the climate action planning process such as elected officials,
environmental and planning nonprofits, advocacy groups, and members
of the public.

We extend our sincere thanks to all those who supported us in this
effort:

Heather Boyer (our editor), Courtney Lix, and the rest of the staff
at Island Press who made this all happen; Dr. Hema Dandekar, Head of
the City and Regional Planning Program at California Polytechnic State
University in San Luis Obispo (Cal Poly), and R. Thomas Jones, Dean
of the College of Architecture and Environmental Design at Cal Poly
for their encouragement and support; all our colleagues in the City and
Regional Planning Department at Cal Poly for help in editing chapters
and their general cheerleading, and three student assistants, Emily Ewer,
Jordan Cowell, and Arianna Allahyar, who helped with our database; Dina
Perkins for her beautiful chapter pictures and enormous patience; our
colleagues in the Sustainability and Climate Change Services team at
PMC for their willingness to collaboratively and enthusiastically tackle
new challenges and set new standards in the field of climate action planning; Ken Yocom, Assistant Professor of Landscape Architecture at University of Washington, and Vivek Shandas, Associate Professor of Urban
Studies and Planning at Portland State University, for their willingness
to review draft chapters; and five anonymous reviewers.

We especially want to thank the following community leaders who
were interviewed for the community cases:

In Portland and Multnomah County, Oregon: Michael Armstrong,
    Senior Sustainability Manager, City of Portland Office of Sus-
    tainability.
In Evanston, Illinois: Elizabeth Tisdahl, Mayor of Evanston; Paige K.
    Finnegan, Chief Operating Officer at e-One, LLC, and Co-Chair

of the Evanston Environment Board; and Dr. Stephen A. Perkins, Senior Vice President, Center for Neighborhood Technology.

In Pittsburgh, Pennsylvania: Lindsay Baxter, City of Pittsburgh Sustainability Coordinator.

In San Carlos, California: Deborah Nelson, Planning Manager, City of San Carlos.

In Miami-Dade County, Florida: Susanne M. Torriente, Sustainability Director (Plan Leader), and Amy Knowles, Organizational Development Administrator, at Miami-Dade County.

In Homer, Alaska: Anne Marie Holen, Special Projects Coordinator, City of Homer.

Chapter One

───────── ✐ ─────────

Climate Action Planning

Global warming is real and demands our immediate response. It is in our national interest to act now and mayors understand that a successful plan in this country for reducing our energy consumption begins in cities and local communities. We are leading by example in the fight against global warming and representing America to the world.[1]

Gregg Nickels, U.S. Conference of Mayors president
and Seattle mayor

The U.S. Global Change Research Program's June 2009 report to the president and Congress clearly establishes the nature of the global warming problem:

> Observations show that warming of the climate is unequivocal. The global warming observed over the past 50 years is due primarily to human-induced emissions of heat-trapping gases. These emissions come mainly from the burning of fossil fuels (coal, oil, and gas), with important contributions from the clearing of forests, agricultural practices, and other activities.[2]

Global warming is already impacting human health and safety, the economy, and ecosystems. As greenhouse gas emissions continue to accumulate in the atmosphere, global warming impacts will increase in severity. The global challenge is twofold: reduce the human-induced emissions of heat-trapping gases, and respond to the negative impacts already being felt and the likelihood that they will worsen in the future.

1

Figure 1.1   2008 CO_2 emissions from fossil fuel combustion by sector and fuel type.

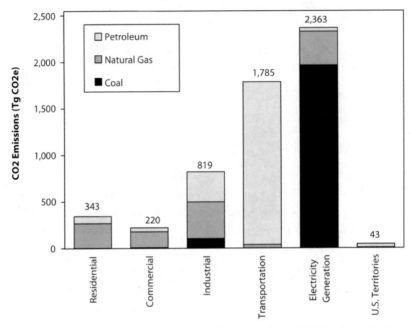

Source: Reprinted from U.S. Environmental Protection Agency, *Inventory of U.S. Greenhouse Gas Emissions and Sinks: 1990–2008*, EPA 430-R-10-006 (Washington, DC, 2010, fig. ES-6).

The largest source of heat-trapping gases, or greenhouse gases, is fossil-fuel-burning power plants, and the second-largest source is fossil-fuel-burning vehicles (fig. 1.1). For the former, changes such as better technology, development of large-scale renewable energy, and retirement of old, inefficient power plants will have an important role to play in reducing greenhouse gas emissions. For the latter, evolving vehicle and fuel technology and standards will help reduce greenhouse gas emissions. These types of technological evolution and large-scale energy programs are driven by private-sector investment and federal and state government legislation and programs. Although these efforts are important and necessary, the problem of global warming cannot be solved without the participation of communities, local governments, and individuals as well.

Local action is critical for needed greenhouse gas emissions reductions to occur. Local governments control the vast majority of building construction, transportation improvements, and land use decisions in the United States. Civic and business organizations, environmental groups, and citizens can join forces with local government and commit to local action that includes energy efficient operation of local government, energy efficient buildings, alternatives to driving such as city buses and bicycles, and city planning that improves the quality of life and allows people to depend less on their car.

Fortunately, communities all over the United States are responding to the challenge of climate change by assessing their greenhouse gas emissions and specifying actions to reduce these emissions. As of early 2011, over a thousand mayors had signed the U.S. Mayors Climate Protection Agreement (box 1.1), vowing to reduce carbon emissions in their cities below 1990 levels, in line with the Kyoto Protocol (fig. 1.2).[3] In October 2009, Mayor Scott Smith of Mesa, Arizona, became the 1,000th signatory to the Agreement. At the signing ceremony, he expressed the needed collective effort: "I welcome the opportunity to join with 1,000 of my peers in this truly bipartisan effort to improve not only the environment, but our communities and our nation. We may not all agree on specific action points, but we are united in a common goal of responsible environmental stewardship."[4] When Mayor Tim Davlin of Springfield, Illinois, signed the Agreement he reminded everyone what it would take and why it was needed: "We must rally the

Figure 1.2   Map showing cities participating in the U.S. Mayors Climate Protection Agreement (as of February 2011).

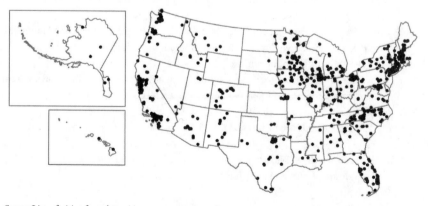

Source: List of cities from http://www.usmayors.org/.

Box 1.1
The U.S. Mayors Climate Protection Agreement[a]

A. We urge the federal government and state governments to enact policies and programs to meet or beat the target of reducing global warming pollution levels to 7% below 1990 levels by 2012, including efforts to: reduce the United States' dependence on fossil fuels and accelerate the development of clean, economical energy resources and fuel-efficient technologies such as conservation, methane recovery for energy generation, waste to energy, wind and solar energy, fuel cells, efficient motor vehicles, and biofuels;

B. We urge the U.S. Congress to pass bipartisan greenhouse gas reduction legislation that (1) includes clear timetables and emissions limits and (2) a flexible, market-based system of tradable allowances among emitting industries; and

C. We will strive to meet or exceed Kyoto Protocol targets for reducing global warming pollution by taking actions in our own operations and communities such as:

1. Inventory global warming emissions in City operations and in the community, set reduction targets and create an action plan.

2. Adopt and enforce land use policies that reduce sprawl, preserve open space, and create compact, walkable urban communities;

3. Promote transportation options such as bicycle trails, commute trip reduction programs, incentives for car pooling and public transit;

4. Increase the use of clean, alternative energy by, for example, investing in "green tags," advocating for the development of renewable energy resources, recovering landfill methane for energy production, and supporting the use of waste to energy technology;

5. Make energy efficiency a priority through building code improvements, retrofitting city facilities with energy efficient lighting and urging employees to conserve energy and save money;

6. Purchase only Energy Star equipment and appliances for City use;

7. Practice and promote sustainable building practices using the U.S. Green Building Council's LEED program or a similar system;

8. Increase the average fuel efficiency of municipal fleet vehicles; reduce the number of vehicles; launch an employee education program including anti-idling messages; convert diesel vehicles to bio-diesel;

9. Evaluate opportunities to increase pump efficiency in water and wastewater systems; recover wastewater treatment methane for energy production;

10. Increase recycling rates in City operations and in the community;

11. Maintain healthy urban forests; promote tree planting to increase shading and to absorb CO_2; and

12. Help educate the public, schools, other jurisdictions, professional associations, business and industry about reducing global warming pollution.

[a] As endorsed by the 73rd Annual U.S. Conference of Mayors meeting, Chicago, 2005.

Figure 1.3   Map showing cities with completed, stand-alone climate action plans (as of February 2011).

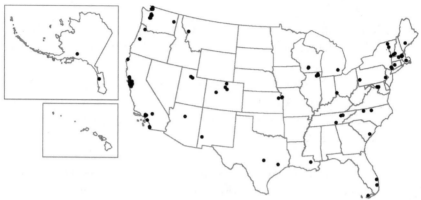

Source: Authors.

entire community to creatively find additional ways to reduce emissions and make our planet a better place to live for our children and their children."[5]

These kinds of commitments have driven the completion of over 120 city and county climate action plans (CAPs) as of early 2011 (fig. 1.3).[6] Most of these are only a few years old so their impact is yet to be felt, but some communities are well into implementation of their greenhouse gas emissions reduction strategies and are beginning to report success. ICLEI–Local Governments for Sustainability (ICLEI),[7] a "membership association of local governments committed to advancing climate protection and sustainable development," in their 2009 annual report notes the following successes (among others):[8]

- Broward County, Florida, reduced emissions by 62,491 metric tons of greenhouse gases annually between 1997 and 2007.
- Portland, Oregon, reduced local carbon emissions in 2008 to 1% below 1990 levels, despite rapid population growth.
- New York City, New York, in September 2008 reported a 2.5% reduction in citywide greenhouse gas emissions between 2005 and 2007, largely due to the impact of new natural gas power plants that came online in 2006.

- San Francisco, California, reduced community-wide emissions by 5% between 1990 and 2005—8% from peak emissions in 2000—totaling 670,000 tons of greenhouse gases.
- Minneapolis, Minnesota, reduced community-wide emissions by 7% (440,700 metric tons) between 2000 and 2006, over 50% of which was due to reductions in electricity usage.
- Seattle, Washington, reduced its greenhouse gas emissions to 8% below the 1990 baseline by 2005.
- Boulder, Colorado, has been reducing community emissions since 2006.

These successes show that actions to reduce greenhouse gas emissions can work and that aggressive reduction targets can be met.

Also at the local level, many U.S. colleges and universities are leading the way in climate action planning. As of early 2011, about 380 U.S. colleges and universities have adopted a CAP, with several hundred more committed to action.[9] There is a great opportunity for communities to partner with their local colleges and universities to share knowledge and resources and engage in collaborative planning.

The tremendous variety of efforts taking place in cities, counties, and colleges and universities to address the problem of climate change is impressive and suggestive of the need to establish "best practices" in this new field of planning for greenhouse gas emissions reduction and for climate change adaptation. Although the specific names for these plans vary, they are generally referred to as climate action plans (CAPs). This book provides basic guidance on preparing a local CAP and making key decisions about methods and assumptions that all plan writers should address. The information in the book should be useful to cities, counties, and colleges and universities since the basic climate action planning process is the same.

What Are Climate Action Plans?

CAPs are strategic plans that establish policies and programs for *reducing* (or mitigating) a community's greenhouse gas (GHG) emissions and *adapting* to the impacts of climate change. CAPs may be visionary, setting broad outlines for future policy development and coordination, or they may be focused on implementation with detailed policy and program information. Although there is no official format or content guide for CAPs, the most commonly used has been ICLEI's *Cities for Climate Protection Milestone Guide.* A review of existing guidance and adopted CAPs shows that they are usually based on GHG emissions inventories and forecasts, which identify the sources of emissions from the community and quantify the amounts. They also identify a GHG emissions reduction goal or target. To reduce emissions and meet the reduction target, CAPs typically focus on land use, transportation, energy use, and waste—since these are the sectors that produce the greatest amount of GHG emissions—and may differentiate between community-wide actions and local government agency actions. This book refers to these actions as *emissions reductions* or *reduction strategies*, rather than using the terms *mitigation* or *mitigation strategies.* Additionally, many CAPs now include a section addressing how the community will respond to the impacts of climate change on the community such as sea level rise, increased flooding, and change in ecological processes; this is usually referred to as *climate adaptation* (box 1.2).

CAPs can be stand-alone documents or they may be integrated into comprehensive land use plans, "green" plans, sustainability plans, or other community-level planning documents (box 1.3). For example, New York City prepared a sustainability plan titled *PlaNYC* that addresses housing, open space, brownfields, and water and air quality, as well as climate change. Some communities may have climate action policies and programs in various documents and resolutions that are collectively the equivalent of a unified CAP. Increasingly though, CAPs are prepared as stand-alone documents. This book focuses on the preparation of stand-alone CAPs and suggests how they can be integrated with other community plans and policies.

CAPs vary in role and content based on community context and local vision. *Role* refers to the functions that the plan performs in the community. *Content* refers to the topics or issues that the plan covers. Communities need to consider the following points as they make decisions about the roles and contents of their own CAPs. In turn, these decisions should direct the climate action planning process.

A CAP performs these functions in a local community:[10]

Box 1.2
Defining Emissions Reduction (Mitigation) and Climate Adaptation

Terminology is not consistent across CAPs or CAP guidance documents. Two common terms are climate *mitigation* and climate *adaptation*; CAPs often address both. This book, rather than referring to *mitigation* or *mitigation strategies*, refers to *emissions reductions* or *reduction strategies* as the preferred terminology. Either terminology refers to actions that reduce the net amount of GHG emissions to the atmosphere.

Climate adaptation refers to actions taken to improve a community's resilience when confronted with impacts of climate change. This usually includes addressing sea level rise, changes in weather and rainfall, and increased susceptibility to natural disasters such as wildfires, floods, and hurricanes. Climate adaptation planning is linked very closely to hazard mitigation planning, and this often creates confusion over terminology. To avoid this confusion, this book uses the terms *climate adaptation* and *adaptation*.

Climate change is like an imminent car crash.
Mitigation is the brakes—it will reduce the magnitude of the impact of climate change.
Adaptation is the airbags—it will soften the blow.
We need BOTH to survive the crash intact.[a]

[a] Geos Institute and Local Government Commission, *Integrated Strategies for a Vibrant and Sustainable Fresno County* (March 2011), 18.

1. Establishes actions necessary to reduce local GHG emissions and meet desired targets
2. Establishes actions for adapting to climate change–induced impacts and hazards
3. Establishes accountability for action
4. Brings stakeholders together
5. Informs the public
6. Integrates actions from various community plans
7. Integrates actions across different scales (local, regional, state, federal, international)
8. Saves money through energy efficiency and builds the local economy
9. Improves community health and livability
10. Responds to local context and conditions

> ## Box 1.3
> ### Types of Local Plans Addressing Climate Change
>
> Communities may choose to address climate change through a variety of local planning documents. The following four are the most common types:
>
> *Climate action plans*: Stand-alone plans specifically addressing climate change issues and based on local greenhouse gas (GHG) emissions inventories.
>
> *Sustainability and "green" plans*: Plans that address a variety of sustainability, "green," or environmental issues but include a climate action section and may be based on a GHG inventory.
>
> *Energy plans*: Plans that focus on energy efficiency and conservation but include a climate action section and may be based on a GHG emissions inventory.
>
> *Comprehensive/general/community plans*: Community land use plans that include an element or sections that address climate action and may be based on a GHG emissions inventory.

The following are standard contents of a CAP (box 1.4):

1. Background on climate change and potential impacts
2. Inventory of local GHG emissions
3. Forecasts of future GHG emissions
4. GHG emissions reduction targets
5. Emissions reduction strategies (quantified and based on the best available science and appropriate for the jurisdiction) that cover energy, transportation, solid waste, and land use
6. Adaptation strategies based on the best available science and appropriate for the jurisdiction
7. Implementation program, including assignment of responsibility, timelines, costs, and financing mechanisms
8. Monitoring and evaluation programs

CAPs have two technical or quantitative components that can make them more challenging to prepare than traditional community-level plans: the GHG emissions inventory and the GHG emissions reduction strategies. The GHG emissions inventory is an identification and accounting of GHGs emitted to the atmosphere from sources

Box 1.4
An Example of a Climate Action Plan Table of Contents

Town of Bedford, New York, Climate Action Plan Table of Contents

I.   Introduction
    a.   Climate Science
    b.   International & National Policy
    c.   Our Climate Action Plan (CAP)
    d.   Sustainable Bedford

II.   Greenhouse Gas Emissions Inventory
    a.   Introduction
    b.   Methodology and Model
    c.   Creating the Inventory
    d.   Community Emissions Inventory
    e.   Municipal Emissions Inventory
    f.   Conclusion

III. Reduction Measures
    a.   Summary
    b.   Energy
      —Municipal
      —Community
    c.   Transportation
      —Municipal
      —Community
    d.   Waste + Recycling
      —Municipal
      —Community
    e.   Land + Water Use
      —Municipal
      —Community

IV.   Implementation
    a.   Bedford 2020 Coalition
    b.   Measuring Our Progress
    c.   Implementation Timeframe Table

V.   Glossary and Acronyms

VI.   Acknowledgements

VII. Sources

VIII. Appendix

Source: Town of Bedford, New York, *Town of Bedford Climate Action Plan*, accessed March 1, 2011, http://www.bedfordny.info/html/pdf/green/2009%20Sept%20Draft%20Action%20Plan.pdf.

within the community over a period of time, usually a calendar year. These emissions are not measured directly; instead they are estimated based on quantifying community activities and behaviors such as the number of miles driven in vehicles and the amount of electricity consumed by residences and businesses. For example, the City of Hamden, Connecticut, conducted a GHG emissions inventory and determined that the community emitted 613,233 metric tons of GHGs in 2001.[11] The emission sources were nearly evenly split among the residential sector (37%), the transportation sector (34%), and the industrial and commercial sector (24%), with a small contribution by the waste sector (5%). The GHG emissions inventory also usually contains projections of future emissions that provide a basis for reduction targets and a benchmark for progress toward achieving them.[12]

There are various approaches for inventorying GHG emissions, but the lack of a clear, consistent protocol has frustrated many local efforts. Fortunately, a recent partnership among ICLEI, the California Air Resources Board, the California Climate Action Registry, and The Climate Registry produced a standard approach useable nationwide for inventorying GHG emissions that result just from local government operations (the Local Government Operations Protocol [LGOP]) such as fueling vehicle fleets and powering government facilities. Additional work by ICLEI is under way to develop a similar protocol for emissions from the whole community and is expected to be available in early 2012.

The complement to the GHG emissions inventory is development of GHG emissions reduction strategies. Reduction strategies are tied quantitatively to the emissions detailed in the inventory to demonstrate a plan's ability to reach emissions reduction targets. Predicting emissions reductions from reduction strategies requires that numerous assumptions be made about future local behavior and feasibility of implementation for each strategy. For example, the City of Cincinnati, Ohio, identified a reduction strategy in collaborating with "regional bicycling advocates in order to increase bicycle use as a mode of transportation." They then estimated that it would reduce annual GHG emissions by 6,300 tons per year by gathering data on existing and forecasted transportation mode share, average bicycle trip length, and vehicle emissions factors. A key assumption was that this collaboration could achieve a fourfold increase in the percentage of workers over the age of 16 that bike to work.[13] Through future monitoring and evaluation the City could determine the accuracy of their predictions and make needed adjustments.

Most CAPs include emissions reduction strategies in the areas of land use, transportation, renewable energy and clean fuels, energy conservation and efficiency, industrial and/or agricultural operations, solid waste management, water and wastewater treatment and conveyance, green infrastructure, and public education and outreach. Although these categories are fairly consistent across plans, the reduction strategies within the categories vary. Climate adaptation strategies also share common categories, such as buildings and infrastructure, human health and safety, economy, and ecosystems with variation among local measures. For a CAP to be implementable, it must reflect the local context, including emissions sources and relative amounts, geographic location, existing policy, employment base, transportation modes, development patterns, community history, and local values and traditions. These factors inform decision making as to which emissions reduction or adaptation strategies are most likely to be locally effective.[14]

CAPs often include a discussion of co-benefits of the various identified emissions reduction and climate adaptation strategies. Co-benefits accrue in addition to the primary climate benefits (fig. 1.4). For example, residential energy efficiency programs often decrease homeowners' power bills, bicycling incentives promote health and recreation, and tree planting improves air quality and community aesthetics. Communities may emphasize co-benefits more than climate benefits to garner broader support for climate action planning. For example, in Salina, Kansas, the Climate and Energy Project has focused on energy efficiency (i.e., saving money) and green job creation;[15] these are touted as the primary benefits, and GHG emissions reductions are seen as co-benefits.

Because climate action planning has novel technical requirements, CAP preparation is becoming a professional activity. In addition to nonprofit organizations such as ICLEI that specialize in providing planning guidance and technical assistance, a number of consulting firms specialize in GHG emissions inventories and climate action planning services. Some communities are creating high-level staff positions to oversee preparation and implementation of climate action and sustainability

Figure 1.4   Categories of co-benefits identified in the draft City of San Luis Obispo (California) Climate Action Plan.

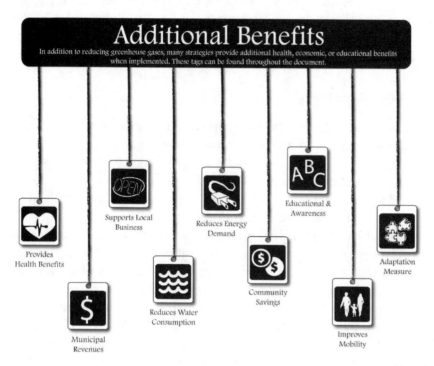

plans. Professional associations are offering training and support for members specializing in climate change issues. Colleges and universities are offering classes and certificate programs, and full-degree programs are emerging. This book contributes to this emerging field by guiding climate action planners, and others interested in the field, through the plan development process by identifying the key considerations and choices that must be made in order to assure a locally relevant, implementable, and effective plan.

Why Is Climate Action Planning Needed?

Climate change is a global phenomenon that cannot be adequately addressed at any one scale. Both reducing GHG emissions and adapting to unavoidable climate impacts require action at the local level as well as the state, federal, and international levels. This section summarizes the science and predicted impacts of climate change globally and in regions of the United States (appendix A provides an in-depth discussion of the science of climate change). Following these descriptions are discussions of the need for solutions at the global and local scales.

The Global Problem

Without an atmosphere and the natural greenhouse effect, Earth's average global temperature would be around freezing. When considered in this context, the greenhouse effect is a physical phenomenon on which human life and civilization and other forms of life as we know it depend. The greenhouse effect is due to the presence of carbon dioxide, water vapor, and a few other chemicals in the atmosphere (i.e., GHGs). In the manner of a greenhouse, these chemicals help trap heat and thus keep Earth's temperature within a life-sustaining range. The problem is that human activities such as burning fossil fuels in power plants and automobiles, clearing tropical forests, and operating modern agricultural systems have produced additional GHGs that are accumulating in the atmosphere and generating additional global warming.

To better understand the nature of this accumulation and its potential impacts, the United Nations Environment Programme (UNEP) and the World Meteorological Organization (WMO) established the Intergovernmental Panel on Climate Change (IPCC) "to provide the world with a clear scientific view on the current state of climate change and its potential environmental and socio-economic consequences."[16] The IPCC is an international group of over a thousand scientists who review and summarize climate science and issue periodic reports. These reports represent the consensus of these scientists as to the best knowledge we have about climate change. The IPCC, in their 2007 reports, state the following:[17]

> Carbon dioxide is the most important anthropogenic [human-caused] greenhouse gas. The global atmospheric concentration of carbon dioxide has increased from a pre-industrial value of about 280 ppm to 379 ppm in 2005 [fig. 1.5]. The atmospheric concentration of carbon dioxide in 2005 exceeds by far the natural range

over the last 650,000 years (180 to 300 ppm) as determined from ice cores. . . . The primary source of the increased atmospheric concentration of carbon dioxide since the pre-industrial period results from fossil fuel use, with land use change providing another significant but smaller contribution.

The IPCC also identifies the sources and levels of other GHGs such as methane and nitrous oxide. They then discuss the combined "radiative forcing" of these GHGs. Radiative forcing is simply the concept that certain forces may change the energy balance of Earth's climate. GHGs create positive radiative forcing and thus can drive a net increase in Earth's temperature. There are negative radiative forcings as well—such as increased cloud formation—but the IPCC has concluded that the positive forcings are larger.

Figure 1.5   Annual global temperature anomaly (NASA Goddard Institute for Space Studies data) and CO_2 levels from ice cores at Law Dome, Antarctica (Oak Ridge National Laboratory Carbon Dioxide Information Analysis Center data) and atmospheric measurements at Mauna Loa, Hawaii, USA (National Oceanic and Atmospheric Administration data).

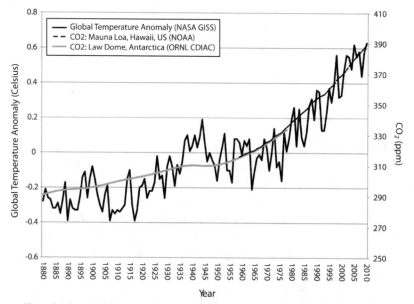

Source: Chart developed based on information from the website Skeptical Science: http://www .skepticalscience.com/co2-temperature-correlation.htm.

The IPCC reached several conclusions about the effects of the positive radiative forcing:

- Warming of the climate system is unequivocal, as is now evident from observations of increases in global average air and ocean temperatures, widespread melting of snow and ice, and rising global average sea level (see fig. 1.5).[18]
- Most of the observed increase in global average temperatures since the mid-twentieth century is very likely due to the observed increase in anthropogenic (human-caused) GHG concentrations.[19]
- For the next two decades, a warming of about 0.2°C per decade is projected.[20]
- Continued GHG emissions at or above current rates would cause further warming and induce many changes in the global climate system during the twenty-first century that would very likely be larger than those observed during the twentieth century.[21]
- A global assessment of data since 1970 has shown it is likely that anthropogenic warming has had a discernible influence on many physical and biological systems.[22]
- Some large-scale climate events have the potential to cause very large impacts, especially after the twenty-first century.[23]

These conclusions demonstrate that there is a global problem in both cause and effect.

The Local Problem

Climate change is a global problem, but the impacts of climate change will be felt locally through disruptions of traditional physical, social, and economic ways of life. In some cases these changes may be positive such as the lengthening of growing seasons in midlatitudes, but in most cases they will be negative. In 2009 the U.S. Global Change Research Program released a report titled *Global Climate Change Impacts in the United States*,[24] which summarizes the impacts of climate change on the United States. The key findings of the report included the following:

- Climate-related changes already observed in the U.S. and projected to grow include
  – increases in heavy downpours

    – rising temperature and sea level
    – rapidly retreating glaciers and thawing permafrost
    – lengthening growing seasons
    – lengthening ice-free seasons in the ocean and on lakes and rivers
    – earlier snowmelt
    – alterations in river flows

- Climate changes are already affecting water, energy, transportation, agriculture, ecosystems, and health.
- Climate change will stress water resources.
- Crop and livestock production will be increasingly challenged.
- Coastal areas are at increasing risk from sea level rise and storm surge.
- Risks to human health will increase, including heat stress, waterborne diseases, poor air quality, extreme weather events, and diseases transmitted by insects and rodents.

These changes will vary regionally so each community will experience a different mix of types and severities of impacts.

Some states have prepared similar analyses of the risks presented by climate change to assist communities in more clearly understanding their own risks. For example, Alaska identified problems such as melting glaciers, rising sea levels, flooding of coastal communities, thawing permafrost, increased storm severity, forest fires, insect infestations, and loss of the subsistence way of life as animal habitat and migration patterns shift and as hunting and fishing become more dangerous with changing sea and river ice.[25] South Carolina is concerned about changes to agriculture, loss of forests, availability of water supplies, and increased electricity demand.[26] In California the combined effects of increased drought, temperatures, wildfires, floods, and sea level rise could result in "tens of billions per year in direct costs, even higher indirect costs, and expose trillions of dollars of assets to collateral risk."[27]

At the local level, cities and counties are identifying their climate risks and developing adaptation strategies. For example, Chicago identified more intense heat waves and responded with changes to the building code, an aggressive tree planting program, and a revised emergency response plan. Miami, Florida, at an average of only 1.8 meters above sea level, has begun long-term planning for infrastructure, flood mitigation, and water supply.[28] Aspen, Colorado, is worried about economic impacts to its world famous ski resorts. The ski season is predicted to start later and end earlier, with a decreased ability to make and maintain snow during the season.[29] For each of these communities the local problem is very real and provides enough justification to act.

The Need for Solutions beyond the Local Level

Some aspects of emissions reduction and climate change adaptation require action at larger scales. The connected nature of materials flows (including global trade), economic markets, and information results in the need for governmental mandates on national and international scales. These policies level the playing field to alleviate the potential loss in competitive advantage by enacting local climate policy. These larger-scale strategies also provide context for local efforts by addressing those emissions sources that fall outside local jurisdictional control.

The most widely recognized of the international efforts to address climate change is the Kyoto Protocol, an international agreement adopted in 1997. The Protocol defined GHG emissions reduction targets and outlined a series of strategies to reach these targets, most of which are market-based mechanisms, including a cap-and-trade or emissions trading system. The emissions reduction target, which varied slightly by participating country, averaged 5% below 1990 emissions levels by 2012. This target has been used as the basis for many local climate planning efforts and was specified in the U.S. Mayors Climate Protection Agreement (see box 1.1). In addition, Kyoto signatories were required to measure or inventory their emissions and identify measures to reduce them.[30] The Kyoto Protocol has resulted in an overall reduction in emissions, but these reductions are not uniformly distributed among participating countries or emissions sectors.[31] Between 1990 and 2007 emissions have dropped in some countries but not all. There is similar variation among emitting sector and type of GHG. Carbon markets have proved effective for curbing manufacturing emissions but have not resulted in reduced emissions from transportation or energy sectors.

In December 2009, the United Nations Climate Change Conference was held in Copenhagen, Denmark. The aim was to define an agreement to guide actions beyond the 2012 target year of the Kyoto Protocol. The resulting Copenhagen Accord is less detailed than the

Kyoto Protocol, but it includes several measures of note. Several countries that support the Accord had not signed the Kyoto Protocol, most notably the United States and China. Instead of an emissions target tied to emissions levels in the past, the Accord sets a goal of keeping the global temperature increase at 2°C or less. The Accord also placed a greater emphasis on the need for adaptive actions in the face of climate change impacts.[32] In February 2010, the countries supporting the Copenhagen Accord submitted their national 2020 reduction targets. These targets vary widely, and many are contingent on the action of national legislative bodies.[33] President Obama pledged a reduction of 4% below 1990 levels by 2020 for the United States, but this awaits congressional ratification.

At the national level in the United States, legislative acts, executive orders, court decisions, and agency rulemaking have defined the nation's climate change policy. Perhaps most notable have been the Environmental Protection Agency's (EPA's) decision to consider carbon dioxide a pollutant to be regulated, new rulemaking on automobile efficiency and gas mileage standards, and the failure of Congress to pass a cap-and-trade bill that would further regulate GHG emissions from big industries and utility providers (box 1.5). On climate adaptation, little federal policy direction exists, but President Obama has created the Interagency Climate Change Adaptation Task Force to provide recommendations on this issue.

Two notable North American efforts at GHG emissions reduction are the Regional Greenhouse Gas Initiative (RGGI) and the Western Climate Initiative (WCI). The RGGI includes ten northeastern and mid-Atlantic states with the aim of reducing GHG emissions from the power sector by establishing a cap to achieve a 10% reduction by 2018. The WCI includes seven western states and four Canadian provinces with a goal to achieve a reduction of 15% below 2005 emissions levels by 2020. WCI addresses a much broader spectrum of emissions sources as compared with the RGGI. Where RGGI sets an emission cap on the energy sector, WCI seeks to cap emissions associated with electricity generation, industry, transportation, and residential and commercial fuel use. Success for WCI participants will rely to a greater degree on local action as well as state-level mandates. RGGI and WCI have not been in operation long enough to assess the degree of success.

Box 1.5
Notable Recent Federal Actions on Clean Energy and Climate Change

Recovery Act Investments in Clean Energy

The American Recovery and Reinvestment Act included more than $80 billion in the generation of renewable energy sources, expanding manufacturing capacity for clean energy technology, advancing vehicle and fuel technologies, and building a bigger, better, smarter electric grid.

Federal Agency Sustainability

An Executive Order on Federal Sustainability commits the federal government to lead by example and reduce greenhouse gas emissions by 28% by 2020, increase energy efficiency, and reduce fleet petroleum consumption.

Efficiency Standards for Cars and Trucks

For the first time the United States will set joint fuel economy/greenhouse gas emissions standards for cars and trucks and create efficiency and emissions standards for medium- and heavy-duty cars and trucks. This is being done under the Environmental Protection Agency's Corporate Average Fuel Economy (CAFE) program.

Making Homes More Energy Efficient

The Recovery Through Retrofit program will eliminate key barriers in the home retrofit industry by providing consumers with access to straightforward information about their home's energy use, promoting innovative financing options to reduce upfront costs, and developing national standards to ensure that workers are qualified and consumers benefit from home retrofits.

Monitoring Emissions

For the first time, the United States will catalogue greenhouse gas emissions from large emission sources—an important initial step toward measurable and transparent reductions.

Climate Change Adaptation

An Executive Order on Federal Leadership in Environmental, Energy, and Economic Performance calls on the Interagency Climate Change Adaptation Task Force to develop federal recommendations for adapting to climate change impacts both domestically and internationally.

Cap-and-Trade Legislation

Although numerous climate bills aimed at creating a national cap-and-trade program for greenhouse gases were introduced in the 111th U.S. Congress—most notably the American Clean Energy and Security Act of 2009 (ACES)—none passed. Future action is uncertain, but this is a significant area of potential future climate action.

Source: Compiled from "Energy and Environment," U.S. White House, accessed February 20, 2011, http://www.whitehouse.gov/issues/energy-and-environment.

Figure 1.6 Map showing states with climate action plans (as of February 2011). States with completed plans shown in gray and with plans in progress shown in hatch.

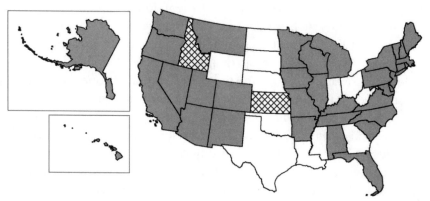

Source: List of states from http://www.pewclimate.org/what_s_being_done/in_the_states/state_action _maps.cfm.

State-level actions are too numerous to list here. As of early 2011, thirty-six states had completed CAPs (with at least two other state plans in progress) (fig. 1.6), and twelve states had begun work on or completed climate adaptation plans (see chap. 6, fig. 6.3).[34] These plans contain a host of emissions reduction and climate adaptation strategies. State-level plans generally set overarching goals but tend to be weak in terms of specific policy action items and are likely less effective than local plans.[35] Nevertheless, communities should consult state plans when developing their own local plans; they may provide information or the opportunity for multiscale coordination of planning.

These international, national, and state-level actions will affect local communities, so it is important that they are considered when developing greenhouse gas emissions inventories and when developing emissions reduction and climate adaptation strategies. Legislative action or government programs that increase fuel efficiency, increase renewable energy, or incentivize energy efficient buildings provide a foundation on which local climate plans may build.

The Need for Local Solutions

Although solutions for climate change are needed at all levels of government, there is a clear need for action at the local level. Globally, cities consume 75% of the world's energy and emit 80% of the GHGs.[36] These emissions come from the cars and trucks we drive; the houses and buildings we heat, cool, and light; the industries we power; and the city services on which we depend. The impacts of climate change will be felt most severely at the local and regional levels as cities become threatened by rising sea levels or increased risk of flood, drought, or wildfire, for example. Changing the way we build and operate our cities can reduce GHG emissions and make cities more resilient against the impacts of climate change.

Several studies have shown the necessity of action at the local level to reduce GHG emissions. In a study of the Puget Sound region of Washington state, researchers determined that an aggressive set of assumptions about future mandated state or national fuel efficiency standards (a 287% increase in fleetwide fuel economy) would still require actions to reduce the number of miles traveled in vehicles (VMT) by 20%.[37] Reducing VMT is largely a function of land use planning and alternative transportation availability, which are mostly controlled by local governments. In another study, researchers showed that to reach needed GHG emissions reductions in the United States, part of the reduction must come from a decrease in "car travel for 2 billion, 30-mpg cars from 10,000 to 5,000 miles per year" and a cut in "carbon emissions by one-fourth in buildings and appliances projected for 2054."[38] These reductions can come from local communities reducing VMT and requiring that new buildings meet strict energy codes and existing buildings be upgraded.

In the United States, local governments have primary control over land use, local transportation systems, and building construction. Each of these areas is a critical component of a CAP. Of course, most communities already have plans that address these issues. For them, climate change is perhaps a new motivation or provides a different approach for doing good community planning.

Regardless of the specific purpose or variation of the CAP a community may prepare, the local CAP and planning process is an acknowledgment of the local responsibility for addressing part of the climate change problem (box 1.6). In fact, one could argue that cities are leading the way in the United States. Federal and state action has been slow to emerge, but many cities are well into implementation of their CAPs and

Box 1.6
"Why Waiting Is Not an Option," from the City of Cambridge, Massachusetts, Climate Protection Plan

Because climate systems are complex and we can't predict the nature and extent of the impacts with certainty, some people advocate delaying action. Unfortunately, waiting to resolve the scientific uncertainties in predicting climate could be disastrous.

To slow and eventually reverse global warming, we must lower the concentration or total amount of greenhouse gases in the atmosphere. This means that not only do we have to lower the rate of greenhouse gas emissions, but we have to reduce the total quantity of emissions until they are lower than the rate at which nature removes carbon from the air. Otherwise, the concentration of carbon dioxide and other GHGs will continue to rise as will temperatures. Currently, the rate of human-made GHG emissions is roughly double the rate of removal. Consequently, emissions must fall by at least half to stabilize GHG concentrations at current levels, and even more to lower the concentration. Scientists indicate that ultimately emissions need to fall to 75 to 85% of current levels.

Waiting to take action is dangerous because of the nature of GHGs. When carbon dioxide emitted by a motor vehicle, building furnace, or power plant enters the atmosphere, it will stay there for a long time—50 to 200 years. This means the warming trend cannot be reversed quickly. The longer the wait, the worse the problem becomes.

While uncertainties in predicting how climate will change in the future may cause scientists to overestimate the impact, there are also uncertainties that may cause them to underestimate the impact. For example, it is unlikely that nature will continue to absorb carbon dioxide at current rates; the latest science suggests it will absorb less as natural systems become saturated, and that several factors limit the ability of plants to take up more CO_2.

This plan proposes that rather than gamble that the scientific community is wrong about climate change, Cambridge take action to reduce emissions by taking advantage of existing technology and resources.

already helping solve the global problem and ultimately their own local problems.

Why Prepare a Local Climate Action Plan?

We offer eight possible reasons for preparing a CAP. Communities may acknowledge all or some of them, and there are likely additional reasons:

- *Global leadership*: Communities acknowledge an ethical commitment as a global citizen to help solve the climate change crisis.
- *Energy efficiency*: Communities want to increase energy efficiency and save money.
- *Green community*: Communities want to create a sustainability or green image for the community, possibly to promote tourism or economic development.
- *State policy*: Communities want to be consistent with state policy direction, sometimes due to incentive programs or looming mandates.
- *Grant funding*: Communities want to gain access to funding that depends on having a CAP, GHG reduction plan, or energy efficiency plan in place.
- *Strategic planning*: Communities take the opportunity to organize disparate sustainability, green, and climate action policies under one document and program for ease of management and implementation.
- *Public awareness*: Communities want to raise public awareness of the climate change issue and build support for more ambitious future efforts.
- *Community resiliency*: Communities that have recognized their vulnerability to the impacts of climate change are seeking greater resilience.

What Is Happening in Climate Action Planning?

Climate action planning is occurring all over the United States in a wide variety of communities. As noted earlier, there are about 120 stand-alone, city and county CAPs with GHG emission inventories that have been formally adopted or received. In addition, many communities (one study estimated 11.8% of U.S. local governments[39]) are engaging in actions to reduce GHG emissions whether they have formally developed a CAP or not. The cities with CAPs are a diverse bunch and defy the typical stereotypes as liberal or green. They range in population size from under 3,000 to over 3.8 million; most are under 100,000. They include cities from all four U.S regions and from thirty of the fifty states, though California is the clear leader, with about one-third of all plans.[40] They are places varying from Homer, Alaska—a small, isolated town on the Kachemak Bay that depends on tourism, commercial fishing, and logging—to Madison, Wisconsin—a midsize city that is the capital of the state and home to the prestigious University of Wisconsin-

Madison—to Chattanooga, Tennessee—a city struggling to recover from significant deindustrialization in the 1980s—and to Los Angeles, California—the second-largest city in the United States and the entertainment capital of the world. There is no typical climate action planning community.

To further illustrate the diversity of climate action planning in the United States, four cases of local climate action planning are presented here, and chapter 8 provides in-depth examinations of six other communities. These communities are very different from each other, but they have all decided to address the problem of climate change through the preparation and adoption of a CAP. They illustrate the kinds of climate action planning under way, the level of innovation occurring in communities, and the range of challenges and opportunities present in communities.

The City of Houston, Texas

Houston is the fourth-largest city in the United States and the world capital of the oil and gas industry.[41] It doesn't seem like the place that would be working to reduce GHG emissions and improve energy efficiency, yet it has emerged as an international leader in doing just that. In 2008, Mayor Bill White approved the Houston Emissions Reduction Plan—their version of a CAP—which identified several strategies that would be undertaken by the larger community and fourteen strategies in the following areas that the city would implement:[42]

- Wind energy
- Facility retrofits with energy savings company financing
- LED traffic signals
- Houston airport system's environmental initiatives
- Citywide lighting retrofit project
- Energy efficient vending machines and vending misers
- LEED [Leadership in Energy and Environmental Design] certification for construction of city buildings
- Combined heat and power system at wastewater treatment facilities
- Fleet use and replacement
- The mayor's hybrid initiative
- Texas emissions reduction plan

- Emerging technology
- Recycling program for all city facilities
- Recycling program for residents

Many of these strategies were already under way, so the plan served to coordinate and give heightened profile to the City's activities.

Houston's efforts to reduce GHG emissions, improve energy efficiency, and save money are working; seven of the fourteen strategies in the plan have been completed. Houston is the top-ranked city in the nation for the municipal purchase of green energy, with 34% of their energy coming from wind power.[43] The EPA identified the city as one of the top in the country for the number of energy efficient buildings, with 133 buildings saving tenants $74 million per year.[44] Houston has leveraged their leadership in GHG emissions reduction to successfully be awarded one of EPA's Climate Showcase Communities grants. The $423,000 grant will be used to reduce transportation-related emissions. According to the City's sustainability director, the money will be used to support alternative forms of transportation such as a bike-share program and expansion of electric car infrastructure.[45]

The Houston CAP is not particularly long or detailed but it does lay a clear path forward with specific objectives, timelines, and implementation mechanisms. Combined with a strong commitment from the mayor's office and support from the community, Houston is helping lead the world in climate action planning.

The City of Stamford, Connecticut

Stamford is a midsize city in the greater New York metropolitan area with a diverse population. In 2003 Mayor Dannel Malloy committed the city to participating in the Cities for Climate Protection Campaign. The city prepared a GHG emissions inventory and in 2005 adopted the Local Action Plan that set reduction targets of 20% below 1998 GHG emissions levels by 2018. The plan addresses both municipal and community-wide reduction strategies focused primarily on energy conservation but also including transportation and smart growth programs.

As of 2009 the city had received over $2.7 million in incentives and grants through Connecticut Light and Power Company to fund about seventy municipal energy efficiency projects, including LED traffic lights, school energy efficiency retrofits, and solar-powered electric vehicle charging stations. These projects are estimated to reduce the city's GHG emissions by about 9,000 tons per year, which is about 70% of their 2018 municipal goal. Moreover, the city is expecting substantial energy cost savings and very short payback periods. For Stamford, reducing GHGs is good for the planet and for the city's fiscal health.

To assist the business sector, the City has created an "Energy Improvement District (EID) for the core of the City, which allows large power users, such as office and apartment buildings, to generate their own economical and energy efficient electricity."[46] The EID is expected to help businesses save money, reduce GHG emissions, earn credits toward LEED green building certification, and increase energy reliability. In addition, the City hopes to use the EID to attract high technology businesses into the downtown.

Stamford created their Local Action Plan with the assistance of an ICLEI intern, a grant from a community foundation, and expertise from community volunteers. To help implement the plan, as well as other community environmental programs, in 2007 the mayor established Sustainable Stanford, a volunteer task force composed of members of the City's business, educational, environmental, and religious communities; City staff; and citizens. The City of Stamford has shown that a CAP can be prepared in a cost-effective manner, and the results can provide the community with environmental and economic benefits.

The City of Key West, Florida

Key West is a small city of about 25,000 people located at the southern tip of Florida. The City adopted the City of Key West CAP in October 2009. The plan itself is mostly visionary and leaves implementation to a more detailed planning document to be prepared in the future. The opening statement of the plan clearly explains why the City chose to act:

> Key West is one of the most vulnerable cities to the effects of climate change. Scientists suggest that escalating greenhouse gas emissions threaten to increase the Earth's temperature and raise sea levels. The City of Key West City Commission, observing high tides already at

street level, has committed to take action here at home and to encourage the rest of the world to do so too.[47]

The climate action planning effort was initiated by the mayor who signed the U.S. Mayors Climate Protection Agreement in 2007. This was quickly followed by city commission resolutions committing to climate action planning and setting GHG emissions reduction targets. The City established several focus groups and committees to ensure the greatest amount of public input. In addition, they convened a topic-specific expert panel, several internal city teams, and an umbrella Climate Action Team to bring together the right people to make the plan happen. Key West has set an example of how people and participation are critical to the climate action planning process.

Like the other example communities, Key West adopted numerous measures to mitigate GHG emissions, although, unlike other example communities, Key West recognized their extreme vulnerability to climate change and addressed how they could adapt to the impacts of climate change as well. The City identified sea-level rise and increasing temperatures as threats to the sustainability of the community. The CAP states the following:

> All areas of planning need to be reexamined through the lens of climate change. The plan needs to address ecologically sensitive land planning, floodplain planning, utility planning, zoning and build-back planning and shoreline hardening. The planning process will include vulnerability assessments and risk assessments, so that a climate resilient community with preparedness goals and preparedness action can be established.[48]

The idea of integrating adaptation planning into all areas of community planning and the concept of resilience are key components of successful climate action planning.

The City of Santa Cruz, California

Santa Cruz is a coastal city of about 55,000 people located an hour's drive south of San Francisco. It is known for its beaches, redwood forests, and the University of California–Santa Cruz. Since the city is in the state of California, it must consider the preparation of its CAP in light of state law and policy. California's mix of legislative acts and governor's executive orders has made local climate action planning nearly mandatory, though the City states that it "did not need a State mandate" to act, and it claims a long history of interest in climate change, including joining ICLEI's Cities for Climate Protection Campaign in 1998.[49] When the City updated its General Plan in 2007, it established a GHG emissions reduction target and a number of goals and policies aimed at achieving the target. At that time the City also established the Climate Action Program and made the decision to hire a climate action coordinator.

Climate action coordinators represent a new profession that is small but growing. Some communities are recognizing the need to designate someone with the oversight and responsibility for implementing the numerous emissions reduction and climate adaptation strategies found in CAPs. Moreover, the increasing number of state and federal laws and grant opportunities make hiring a coordinator a prudent choice. In Santa Cruz the climate action coordinator has been tasked with the following responsibilities:[50]

- Conduct the GHG emissions inventories
- Facilitate the completion of the CAP
- Research municipal best practices in reducing GHG emissions and respond to climate change impacts
- Coordinate volunteer and consultant resources
- Coordinate City participation in regional climate change initiatives
- Support internal City staff efforts to reduce and respond to climate change
- Draft and evaluate proposed general action plan programs
- Communicate the City's climate change efforts and initiatives

In early 2011, the City reviewed a draft CAP that would "meet State land use requirements pertaining to climate change, achieve the policies identified in the draft General Plan 2030 update, and accomplish the reduction goals set by City Council."[51] The plan acknowledges there is no silver bullet and focuses on three areas of action: conservation, sustainable lifestyle choices, and renewable energy alternatives. Mayor of Santa Cruz Mike Rotkin summed up the tangible benefits of the City's plan: "We're doing it because it's the right thing to do, and because it will have a positive economic effect. . . . The only growing sector of the economy is the green sector. We are confident that is our future."[52]

Topics Covered in This Book

This book describes the process and methods for preparation of CAPs for local governments. The book is a practical, how-to guide that directs the reader through the principal steps and critical considerations for developing a local CAP. Each chapter concludes with a list of resources on climate action planning. In addition, given the relative newness of climate action planning and the paucity of standard or best-management practices, we address policy development. The treatment of policy development ranges from the assessment of the global implications of climate change to the specific role of local government in addressing the issue. In the book, we advance the theory that the best CAPs are based on sound science, public education and outreach, recognition of global context and external constraints, and integration with existing policies and programs.

Chapter 2 lays out a program for getting started on climate action planning, including issues of community commitment and partnerships, costs and timing, staffing, creating a Climate Action Team, and auditing existing policies and programs. Chapter 3 establishes principles and practices for developing community participation methods, including the important task of educating the public about this new and challenging public policy issue. Chapter 4 describes best practices for inventorying GHG emissions and includes advice on choosing software, acquiring and managing data, developing forecasts, and establishing emissions reduction targets. Chapter 5 focuses on the core of any CAP: GHG emissions reduction strategies. This chapter explains a process for identifying and evaluating measures to reduce the amount of GHGs the community is emitting. The chapter includes numerous examples of how tailoring measures to particular community contexts and capabilities is the key to successful implementation. Chapter 6 addresses climate adaptation and shows the link between this and local hazard mitigation planning. Chapter 7 provides guidance on successful plan implementation, including timing and financing of measures and monitoring outcomes. Chapter 8 discusses six case studies that show how communities have put this all together to develop effective CAPs. Chapter 9 presents our closing thoughts about the potential for local climate action planning to positively transform the way we live, work, and play. Finally, two appendices address the science of climate change and public participation.

Chapter 2

———————— ✦ ————————

Getting Started

Getting started can be the most challenging step of any new planning process. Although climate action planning includes many of the traditional steps in a comprehensive planning process, it presents a set of challenges distinct from other types of plan development. The climate action plan (CAP) requires not only identification of GHG emissions sources and reduction strategies, but also a quantification of their magnitude and a forecast of future change. If the plan addresses climate adaptation, it should include a local vulnerability assessment. Plan development requires technical expertise and detailed data from a variety of sources not usually drawn upon for other types of local plans, which places an added informational and organizational burden on planning efforts.

It is important that a community complete several preliminary steps in the climate action planning process before working on the core components. These preliminary steps warrant special consideration as they will serve as the foundation to the overall climate action planning process. Communities can usually initiate and complete these steps without external assistance from technical experts or consultants. The steps presented in this book are based on traditional steps for comprehensive planning and include observations from a review of CAPs and evolving best practices. They are primarily written for the perspective of a local government leading the climate action planning effort but are readily transferable to other organizations that may be leading the effort.

The most commonly referenced best practice for the climate action planning process is ICLEI's "The Five Milestone Process."[1] The milestones establish a process to guide communities on how to identify and reduce local greenhouse gas (GHG) emissions. ICLEI's five milestone process is as follows:

1.   Conduct a baseline emissions inventory and forecast
2.   Adopt an emissions reduction target
3.   Develop a local CAP
4.   Implement policies and measures
5.   Monitor and verify results

While ICLEI's milestone approach identifies core steps in CAP development, there are several embedded or additional steps that are critical to the CAP process.

This book proposes a three-phase climate action planning process based on the authors' experience and that of others in preparing CAPs: phase I: Preliminary Activities; phase II: CAP Development; and phase III: Implementation and Monitoring. Although the steps are presented in numerical order, many of them overlap or are iterative; thus they should be applied as general organizing principles rather than a stepwise "cookbook" for planning. If a community is addressing climate adaptation in its plans, then step 8 should include a local vulnerability assessment and step 11 should include climate adaptation strategies, as addressed in chapter 6. When moving through these steps it is important to adhere to principles for a good planning process such as transparency and documentation, participation, justification, and consistency. This chapter describes phase I, phase II is described in chapters 4, 5, and 6, and phase III is described in chapter 7.

Climate Action Planning Process

Phase I: Preliminary Activities
  1.   Establish community commitment
  2.   Build community partnerships
  3.   Establish the role of the plan
  4.   Assemble a Climate Action Team (CAT)
  5.   Consider the logistics of plan development
  6.   Establish a public education and outreach program
  7.   Audit existing community policies and programs
Phase II: CAP Development
  8.   Conduct a baseline GHG emissions inventory and interim forecast
  9.   Formulate plan vision and goals

10. Identify a GHG emissions reduction target
11. Develop, evaluate, and specify GHG emissions reduction strategies
12. Quantify GHG emissions reduction strategies

Phase III: Implementation and Monitoring

13. Develop and administer an implementation program
14. Monitor and evaluate implementation and reduction target attainment
15. Modify and update the plan

In phase I (Preliminary Activities), the community establishes a commitment to climate action planning; builds community partnerships; articulates the intended role of the CAP; makes logistical choices such as identifying a funding source, a timeline for plan development, and a CAP author (e.g., city, consultant, stakeholder, task force); develops a CAT; develops a public outreach and education program; and conducts an audit of existing community policies and programs. The order of these preliminary tasks is not critical and will vary based on community needs. For example, in some communities political commitment may be secured after partnerships have been formed. In others, time and funding considerations may affect the type of plan prepared.

In phase II (CAP Development), the community conducts a baseline GHG emissions inventory; develops an interim forecast of future GHG emissions; establishes a vision, goals, and a GHG emissions reduction target; and develops, evaluates, and quantifies GHG emissions reduction. This phase is usually iterative; often the forecast is adjusted based on the policy audit, and the reduction target may be adjusted as the community evaluates potential reduction strategies. Also, in Phase II, the community should conduct a climate change vulnerability assessment and prepare climate adaptation strategies to reduce the community's vulnerability and increase its resiliency. If the community has a Local Hazard Mitigation Plan prepared under the federal Disaster Mitigation Act of 2000 (DMA 2000) or if they have a safety or natural hazards element of a comprehensive plan, they may be able to use this as a starting point since these plans include vulnerability assessments.

In phase III (Implementation and Monitoring) the community develops and administers an implementation program, implements the adopted policies and strategies, monitors and evaluates implementation, assesses whether the GHG emissions reduction target is being attained, and then modifies and updates the plan based on the evaluation and the changing policy environment. The last two steps are critical but often overlooked. Since a CAP establishes a clear, specific numeric target for GHG emissions reduction, accounting for plan success is a relatively straightforward endeavor. CAPs should explicitly establish how this will be done, and plan stakeholders should commit to making needed changes and updates.

These three phases of CAP development reinforce each other. Given this is a new area of planning, communities need to be willing to experiment, innovate, change course, admit failures, and promote successes. The freedom to develop and implement aggressive, innovative emissions reduction and climate adaptation strategies in phase II relies on the strength of the organizational steps taken in phase I and feedback loop provided by phase III. Experimentation and innovation is only possible with careful monitoring and a firm commitment to revise and adapt strategies based on observed effectiveness.

Chapters 4, 5, and 6 address phase II, and chapter 7 addresses phase III tasks in further detail. This chapter focuses on phase I tasks, except public education and outreach, which are covered in chapter 3. These tasks are intended to address the following questions that every community must answer based on its goals, budget, size and characteristics, political climate, and governance structure:

- How will the planning process be structured?
- Who will prepare and implement the plan?
- What should the plan do?
- Who will participate in the planning process?
- How much will the plan cost and how long will it take?
- What is the community already doing to address climate change?

Establish Community Commitment

Chapter 1 outlined several reasons why a community may want to pursue a local CAP. These include the critical nature of the global climate change problem and the need for immediate action. Some communities act out of self-preservation, some from external mandates, and some due to a sense of responsibility to the global community. Regardless of the reason, the commitment to climate action planning should be established through a formal mechanism. Many communities have accomplished this by the mayor signing the U.S. Mayors Climate Protection Agreement and joining ICLEI, by elected officials passing resolutions, or by community leaders issuing proclamations supporting the commitment to addressing climate change. These are all positive steps for a community, but it is important that they move beyond symbolic gestures and rhetoric to specific action.

In addition to formally committing to address climate change, two steps are necessary to move forward with planning. First, the local government must establish work program priorities and then commit staffing, funding, and resources to climate action planning. Whether the local government is leading the planning effort or not, its commitment is necessary for creating a complete CAP. The second is to secure formal commitments of funding, technical expertise, or political support from relevant private and nonprofit organizations. The cooperation and coordination of local government and community partners constitute a successful formula that has been used by most communities engaged in climate action planning.

Build Community Partnerships

Communities preparing a CAP should consider developing partnerships with entities such as government agencies, community associations and nonprofits, colleges and universities, and neighboring communities (table 2.1). Partners can help with data collection, community education and outreach, stakeholder mobilization, implementation, and monitoring (box 2.1). They may also help reduce the cost of plan preparation by donating volunteer hours, providing needed expertise on specific issues, and enhancing the effectiveness of implementation. A successful planning process built on partnerships can also increase the visibility and credibility of a CAP. This helps with implementation in the community.

When identifying potential partnerships, communities should look to existing planning or implementation partnerships as a starting point. For example, many cities already have formal partnerships among public, private, and nonprofit transit, housing, and social service pro-

Table 2.1   Potential community partners for climate action planning

Government agencies	Community associations and nonprofits	Colleges and universities	Neighboring communities
• Regional and environmental, transportation, and planning agencies • Public health agencies	• Chambers of commerce (representing local employers) • Builders associations • Realtors • Environmental groups • Homeowners associations • Green building groups • Bike and pedestrian advocacy groups • Private utilities	• City planning depts. • Architecture and landscape architecture departments • Public policy and administration departments • Geography departments • Business departments • Agricultural departments • Engineering departments • Science departments	• Cities • Counties • Townships • Indian tribes • Special districts and authorities • Military bases • Federal lands • State lands

viders. Also, some communities participate in county and regional co-operative efforts on transportation planning and funding, emergency management and hazard mitigation, air quality, and stormwater management. These partnerships can be leveraged to assist in CAP preparation and implementation. They also assure compatibility of the CAP with other local and regional efforts. This consistency increases the likelihood of successful CAP implementation.

Once potential partners are identified it is reasonable to consider the pros and cons of the partnership and ask the following questions:

- What is their reputation?
- What experiences have other groups had working with them?
- How may they affect the legitimacy or respectability of the planning process?
- What resources do they bring—knowledge, data, money, labor?

Box 2.1
Partnership Examples

The City of Aspen (Colorado) created a community partnership called the Aspen Global Warming Alliance, which includes such organizations as the Aspen Institute, the Aspen Global Change Institute, The Aspen Skiing Company, the Rocky Mountain Institute, Holy Cross Energy, and the New Century Transportation Foundation, among others. The partnership provided guidance and input on creation of the plan and has an ongoing responsibility to assist the city with implementing the plan.

In the city of Chattanooga (Tennessee), the mayor established the Chattanooga Green Committee composed of representatives from industry and construction associations, environmental and green building groups, universities, and government agencies. The committee conducted public education and outreach activities, prepared the CAP, and continues to assist with implementation.

The Town of Brattleboro (Massachusetts) has partnered with the local nonprofit group Brattleboro Climate Protection to implement the CAP by assisting residents, businesses, and town government in energy conservation and conversion to renewable energy sources.

The City of Benicia (California) worked with students and faculty from City and Regional Planning Community Planning Studio at California Polytechnic State University (Cal Poly), San Luis Obispo, to prepare their CAP. Students audited existing city programs and activities, conducted public education and outreach, and proposed measures for reducing the communities' GHG emissions. The Cal Poly team delivered a draft plan to the city, which was subsequently refined and adopted by the city council. The CAP won an academic merit award from the American Planning Association California Chapter.

In San Luis Obispo County (California) the San Luis Obispo Air Pollution Control District (APCD), a regional air quality planning agency, convened the Climate Change Greenhouse Gas Emission Reduction Stakeholder Group for local agency planners and staff. The group meets bimonthly to identify opportunities and share resources to assist their agencies with compliance with state GHG emission reduction guidelines and regulations.

Choosing the Type of Partnership

Partnerships in CAP development must define a plan "owner." The owner assumes primary responsibility for development and implementation of the CAP. This role can be filled by local government or a nongovernment, community-based organization and may be shared or divided between CAP development and CAP implementation. The advantages of a "government-owned" plan are that local governments have regulatory and taxing authority and usually the legitimacy to successfully implement a plan. The disadvantage is that the CAP process may become enmeshed in local government politics, may be superseded by other local priorities, or may succumb to fiscal pressures. The advantages of a "nongovernment-owned" plan are the potential to build strong grassroots support and relieve the CAP of potential legal or political complications. The disadvantages are the lack of a clear implementing authority and the potential for ownership conflicts between disparate community organizations. Regardless of this choice, local governments and community-based organizations should seek each other out, settle the ownership issue, and commit to a partnership.

In many communities, umbrella groups form around the issue of climate action planning. These may be government-appointed task forces or committees (mayor-appointed groups are common) or self-organizing. They may be focused on technical expertise in areas such as climate science, city planning, alternative transportation, energy efficiency and renewable energy, public health, emergency management, and finance. Or they may be focused on bringing together diverse stakeholders within the community such as environmentalists, business and industry representatives, energy and utility providers, developers and builders, alternative transportation advocates, and homeowners associations.

The role of each partner should be established early in the process to avoid confusion, duplication, and turf battles. There are numerous roles to play that can be clarified by asking the following questions:

- Will partners have an advisory role? In this case their role is to assist in developing the plan and review and comment on plan proposals.
- Will partners have an oversight role? In this case their role is to critically review proposals, make decisions, and provide a final endorsement of the CAP.
- Will partners provide technical or implementation assistance? Partners may have specialized knowledge in an area critical to preparation of the CAP or they may have experience in implementing community programs.
- Will partners provide funding? Partners may have funding available for CAP preparation or implementation.
- Will partners conduct education and outreach efforts? Partners can use their networks, memberships, and community standing to provide edu-

cation and outreach for the CAP, both during development and during implementation.

Partnerships with Community and Nonprofit Organizations

Community and nonprofit organizations often partner with government agencies to prepare and implement CAPs. Community organizations may include nonprofits, advocacy groups, foundations, and business associations. These organizations often fill a critical role in CAP development because many of the GHG emissions reduction strategies in a CAP rely, at least in part, on behavior change. Behavior change, such as increased bicycle ridership, reflects overall community awareness and acceptance of alternatives. Close alliance with or support from key local organizations can be critical to building community support for such changes in daily patterns. During plan formulation, carefully selected local organizations are in an ideal position to provide feedback on strategies most likely to be effective. They can assist with outreach and communication to people who might not normally participate in community planning or who may not be aware of climate change issues. In addition, many partners are well positioned to aid in the outreach programs that assure long-term implementation.

Partnerships with Colleges and Universities

Since many colleges and universities have prepared CAPs, communities should check in with them to see if the plans can be coordinated. Colleges and universities can serve as sources of information and provide technical assistance in preparing GHG emissions inventories and CAPs. The GHG inventory for the City and County of Denver (Colorado) was prepared by a faculty and student team from the Department of Civil Engineering at the University of Colorado, Denver. Similarly, the City of Pittsburgh's (Pennsylvania) GHG inventory was prepared by the Heinz School Research Team at Carnegie Mellon with support from a variety of public and private donors. In the city of San Luis Obispo (California), students from the City and Regional Planning Department at California Polytechnic State University (Cal Poly) prepared an initial draft of the city's CAP.

Partnerships with Neighboring Communities

Partnering with neighboring communities presents unique challenges and opportunities. Communities can differ in politics, priorities, demographics, wealth, size, government structure and capacity, and a variety of other factors. But collaborating offers an opportunity to share resources, save money, and coordinate on difficult regional issues. In San Luis Obispo County (California), the county and cities are partnering to share technical information, work with the regional air pollution control district, and develop CAPs that coordinate county-scale climate actions. In south Florida, the counties of Miami-Dade, Monroe, Palm Beach, and Broward have partnered to form The Southeast Florida Regional Climate Change Compact with the purpose of preparing a regional CAP.

Pros of Partnership with Other Communities
- Sharing of knowledge and resources
- Potential to save money through efficiencies in plan development and implementation
- Coordination of actions that address intercity and regional issues such as transportation

Cons of Partnership with Other Communities
- Dependence on other jurisdictions and loss of some control
- Inconsistency in vision and policy direction

A recent study of local hazard mitigation planning—where jurisdictions had the option of preparing a multijurisdictional plan or going on their own and preparing a single jurisdiction plan—showed that 79% of jurisdictions chose the multijurisdictional option.[2] Almost all of them expressed satisfaction with this choice; of these 82% expressed their belief in the effectiveness of a regional approach and 71% cited cost and time savings (efficiency) as motivations. It is likely that these same benefits can be captured in climate action planning.

Establish the Role of the Plan

A CAP is often a stand-alone document; however, some jurisdictions have chosen not to develop a separate plan, but to integrate climate action strategies into their comprehensive land use plan, sustainability plan, or other planning documents. This can vary from integrating climate action strategies throughout existing chapters, to developing a climate change chapter, to creating an appendix or addendum. All of these choices carry different benefits and challenges, meaning the status of the CAP and its position relative to other local plans must be made based on local context, including existing policy, political climate, and the time and funding available for plan development.

As climate action planning becomes more common, communities are considering whether to integrate the CAP with their comprehensive land use plan (also called general plans, city master plans). An update of a comprehensive plan can be a lengthy process when compared to development of a stand-alone CAP. If a jurisdiction would like to ultimately integrate climate policy into its comprehensive plan but does not have the time or funding to do so in the short term, a CAP can be developed that specifically identifies areas of the comprehensive plan that should be revised in the future. This does not imply that a community must wait to implement emissions reduction or climate adaptation strategies until the appropriate sections of a comprehensive plan are due for revision. Many of the strategies common to a comprehensive plan already serve to reduce GHG emissions and foster community resilience. A CAP can build on these strategies, particularly in the short term. In communities where climate action policies have already been integrated into the comprehensive plan, a CAP can serve as an implementation plan. In this case, the relationship between these two plans is similar to that between a comprehensive plan and a zoning code, for example. The CAP would serve to guide detailed implementation of broad principles contained in the comprehensive plan.

In addition to the issues of whether to create a stand-alone plan is the decision of whether to prepare a plan that addresses only local government (i.e., municipal) operations or whether to prepare a plan that is community wide. Local government operations CAPs (also called municipal CAPs) only address those things that local governments have direct control over such as public buildings, government vehicle fleets, public transit, and water and sewer infrastructure. Although local government operations CAPs can be a great way to get started and can serve as an example to the community, they only address a very small percentage of a community's total emissions (typically 3% to 8%). This book assumes, and advocates for, the preparation of a community-wide

CAP that addresses both local government operations and community-wide emission sources such as residential and commercial energy use, private vehicle use, and industrial and agricultural operations.

The role of CAPs can vary in two important ways. First, CAPs can range from being broad, visionary documents that set an overarching frame for climate action or they can be focused, implementation-oriented plans that contain detailed emissions reduction and climate adaptation strategies. Second, CAPs can range from being innovative plans that direct significant change to more modest plans that consolidate and unify a community's existing policies and programs in one place.

In communities that have not directly addressed climate change from a policy perspective or have few existing strategies that reduce GHG emissions, a CAP can be viewed as a way for local jurisdictions to take an initial step. In this situation, a CAP serves a dual purpose: (1) set the local trajectory for future policy development and local planning updates and (2) identify short-term actions that fit within existing policy. In communities that have a large number of preexisting climate-friendly policies (e.g., green building ordinance, renewable energy program, alternative transportation plans, etc.), the CAP can serve as a unifying document for seemingly disparate policies and programs. Such a unifying document provides a link between existing policies and assures that future programs are complementary to those already in place and further the overarching goals of emissions reduction and climate adaptation. The City of Chicago is an excellent example of a CAP that unifies preexisting policy, identifies new strategies, and will serve to guide future action. Prior to Chicago's adoption of its CAP in 2009, the city already had in place a green building agenda, an aggressive tree planting program, and a green roof program.[3,4] The CAP serves as a unifying document for these efforts and a framework for implementing new complementary programs. Moreover, the success of these strategies can be monitored and evaluated since the CAP links them to a specific GHG emissions reduction target.

Based on the various considerations for the role of the CAP, a few typical configurations have emerged:

- *A CAP as a unifying document:* A community's lack of a CAP does not necessarily indicate a lack of action. Many cities have taken aggressive action focused on emissions reduction and climate adaptation through a suite of independently adopted policies. For example, a community does not need a CAP to have a green building ordinance or renewable energy program. In cases where cities have an array of existing climate-related local policy, the CAP can be viewed as a unifying document. In this case, a CAP can bring together existing policies under an overarching community goal and guide the development of future policy.

- *A CAP as a new policy direction.* In direct contrast to cities that fit in the preceding description, some cities may have no adopted plans or policies that directly address GHG emissions reductions and climate adaptation. In these cases, a CAP serves to identify a new policy direction by identifying overarching emissions reduction and climate adaptation goals, policy focus areas requiring feasibility assessment, and extensive education and outreach to the community to build support for future policy development such as integration into comprehensive plan updates.

- *A CAP as a subsection of a larger sustainability effort.* A climate plan, when viewed in the larger context of environmental policy, is narrow in focus. Prior to the emergence of climate action planning, many cities pursued sustainability goals. CAP development can be viewed as one aspect of an overarching sustainability program.

- *A CAP as an additional section or component of a comprehensive plan.* An implementable CAP must be consistent with other local policy. Many strategies in a CAP directly impact building codes, land use patterns, and circulation. The integration of climate-related strategies with local policy such as a comprehensive plan can occur in a variety of ways. Climate change can be identified as an additional component of a comprehensive plan, but perhaps more effective is to time development of a CAP to match that of comprehensive plan updates to assure consistency with overarching goals being incorporated into the comprehensive plan and the implementation of these goals left to a CAP.

Assemble a Climate Action Team (CAT)

Climate action planning is a data intensive planning process that relies on a number of government agencies or departments and organizations, many of which are unaccustomed to being directly involved in the planning process. Regardless of who is identified to prepare the GHG emissions inventory and CAP—either local government staff members, consultants, or community organizations—one of the first steps in the climate action planning process is establishment of a forum for interactions between the CAP authors and other local government staff members. The ease and accuracy of the GHG emissions inventory and the CAP implementation will rely, in part, on the quality and continuity of the collaboration between plan authors and staff members.

This collaboration can be developed through establishment of a CAT. The CAT plays two critical roles in the climate action planning process. The first is provision of data needed to complete the GHG emissions inventory. With completion of the emissions inventory, the role of the CAT changes to advising GHG emission reduction strategy development, assessing feasibility, and, in the long-term, implementing the chosen strategies. Thus the CAT serves in both a technical capacity and a policy capacity.

Through a CAT, tasks such as staff education, data collection, operational documentation, and long-term plan implementation and monitoring are completed. Because establishment of a CAT is an integral part of plan development and implementation, it is critical to assemble the team strategically. This section details CAT formation and the initial steps necessary to lay the groundwork for conducting a GHG emissions inventory and subsequent CAP development and implementation.

Team Members

A CAT consists primarily of government staff from a variety of departments that oversee day-to-day operations and activities, including fleet management, accounts payable and contracts, parks and recreation maintenance, facility management, building permit approval, and long-range planning. Individual members must be well integrated into their respective departments to facilitate and monitor data collection and review CAP materials. Members of a CAT should have some combination of the following characteristics or knowledge areas: (1) familiarity with department operations that will allow for easy identification and collection of the data needed for the GHG emissions inventory, (2) knowledge of department operations and budget procedures to evaluate GHG emissions reduction strategies, and (3) the authority to implement strategies identified in the CAP. In a given department, this may require several people.

Identifying departments for inclusion on the CAT begins with identification of needed information and responsibilities. Table 2.2 is a partial list of information to aid government agencies in identifying departments and personnel for the CAT. Based on local context additional staff and information may be required, such as jurisdictions that include an airport. Staff may not be able to provide all information needed for the emissions inventory. In some cases, a local agency may choose to include members from a partner organization identified as integral to plan development.

The size of the CAT should be limited (e.g., fewer than twenty members) to assure that the team can foster open dialogue and timely review and response to requests. All staff members who will participate in the CAP development and implementation need not be on the CAT. The CAT should include staff members best able to transmit information to colleagues, identify departmental information sources, and have an overall understanding of department operation.

In most departments, the best initial point of contact is the director or manager. The director is able to oversee long-term implementation of strategies, and in the short term, is best positioned to identify the staff who are able to provide the data needed for an emissions inventory. Involvement of department directors or managers also helps ensure the cooperation of all staff within a given department. As the process evolves, staff below the director may play a more direct role in generating data and disclosing operational procedures.

While local government organization structures vary, the following list includes some of the key departments that should be participants in a CAT. These departments are critical not only to the development of a GHG emissions inventory but to long-term implementation of emissions reduction and climate adaptation strategies. This list can be tailored based on government agency size and function, operational control, and organizational structure.

Utilities

Most jurisdictions provide all or a portion of basic services such as water, power, and solid waste services through a utilities department, making this department a critical member of the CAT. A utilities

Table 2.2. Needed expertise of a climate action team

Category	Access to data for greenhouse gas emissions inventory	Knowledge of local government policies and operations
Facilities	Energy use (electricity, natural gas) Year built Square footage Number of employees Hours of operation Traffic signal energy use Quantity, location, bulb type, and energy use of streetlight, parking lot lighting, security lighting	Operational procedures Planned and completed energy or water efficiency upgrades Facilities proposed for closure or construction
Government fleet (including police, fire, transit, general vehicle pool)	Miles traveled Gallons of fuel used Vehicles by make/model/year Refrigerant use/maintenance	Maintenance schedule Fleet replacement/ conversion schedule
Employee travel behavior	Daily commute distance Business travel type and mileage	Daily commute distance Employee commute reduction programs
Transportation	Vehicle miles traveled on local streets Traffic signal and street lamp energy use	Transportation infrastructure design guidelines and maintenance Long-term planning (all modes including bicycle and pedestrian infrastructure)
Water and wastewater	Treatment and conveyance energy use Volume treated and conveyed	Facilities proposed for closure or construction Pump, blower, and lift station efficiency
Solid waste	Volume and/or weight delivered to solid waste facility Disposal associated emissions (e.g., landfill methane production) Transport distance	Local diversion rate Existing waste diversion program effectiveness
Parks and recreation	Fuel type and amount for maintenance equipment (mowers, blowers, etc.) Size of area maintained (i.e., park and open space acreage) Water use	Open space and park area and use Maintenance schedule Irrigation infrastructure type Urban forest management Recreational program administration

Administration/finance	Invoices for vendors related to refrigerant replacement, waste haulers, and others as needed Lists of equipment and vehicles Mileage and destinations for employee travel	Cost feasibility evaluation Budget Capital improvement plan/ program
Long-range planning	Buildout year or horizon year for the comprehensive plan Baseline year of the comprehensive plan Planning area and/or expansion area included in the comprehensive plan	Planned future development Comprehensive plan build out assumptions Existing policy consistency
Development review		Building and project permit approval process
Economic development		Identification of economic constraints and opportunities

Source: Adapted from ICLEI, *Cities for Climate Protection: Milestone Guide* (Oakland, CA: Author, n.d.); California Air Resources Board, *Local Government Operations Protocol for the Quantification and Reporting of Greenhouse Gas Emissions Inventories Version 1.0* (Sacramento, CA: Author, 2008).

department is often the data source for energy use (electricity and natural gas) and operational procedures. This includes the building operations, the treatment and conveyance of drinking water and wastewater, traffic signals and street lights, and solid waste generation. CAT members from a utilities department should be aware of operational changes such as the use of motion detectors or thermostat regulation that were implemented as cost-saving measures. If there is a community-owned utility for power, then the department should have access to data on community-wide use of electricity. Other information useful to the CAT is the success of existing programs such as recycling and educational programs.

Transportation and Engineering (Public Works)
In the United States, transportation-related emissions are the single largest contributor to GHG emissions.[5] The transportation department is critical to assembling accurate data for the emissions inventory, particularly estimates of vehicle miles traveled (VMT) on local streets. In addition, transportation CAT members should be able to provide VMT,

fuel use, and cost information for local government fleet vehicles. In the long-term, CAT members should have the authority to change vehicle purchasing procedures to more fuel-efficient models. They will also be key in developing and implementing strategies for the community's transportation infrastructure and management.

Community Development (Planning and Building)

The community development or planning and building departments must be included on the CAT. Depending on who is developing the inventory and plan, community development staff may be tasked with CAT coordination, assuring that there are no gaps in data or implementation, and overseeing information-gathering efforts that span multiple departments such as employee commute data. Community development staff is also best positioned to aid in the development of the policy audit described later in this chapter. In addition to a coordination role, this department oversees update and implementation of a community's comprehensive land use plan and zoning codes, which should be linked with the CAP. Also, the community development department often houses development review and permitting functions. It is through this process that strategies such as impact fee, energy efficiency, or green building programs may be implemented; therefore, community development is a critical adviser in development of these and similar strategies.

Parks and Recreation

Parks and recreation departments often maintain a vehicle fleet, operations and maintenance equipment, and parks and recreation facilities. The parks and recreation staff are often charged with maintenance of community green spaces such as parks, open spaces, and vegetated areas in the streetscape. Parks and recreation staff help assess fuel, water, and energy efficiency and conservation practices. They also identify opportunities for GHG emission reductions from parks and open spaces, provide sequestration through tree planting, enable local food access through community gardens, and identify potential sites for local alternative energy generation.

Partner Organizations

Depending on jurisdiction size, services (e.g., water, waste, transit) may be provided by outside suppliers. In this case, data and long-term implementation will require the involvement of staff from regional providers or agencies. In addition, partner organizations may have particular expertise that will strengthen plan elements. Where deemed appropriate, a representative from these organizations can be invited to the CAT.

Role of the Team

The CAT's primary role is to contribute to the GHG emissions inventory, policy audit, and CAP development and implementation. Each of these tasks is detailed in a subsequent chapter. However, prior to beginning the climate planning tasks there is a series of educational steps intended to prepare staff for the CAP development process. This educational process is ongoing and iterative. It begins as soon as CAT members are identified and recruited. Potential members may be skeptical of new policy relating to climate change or may feel uncomfortable with increased demands on staff time. The formality required for this initial outreach varies. A department that is hesitant to participate requires a greater effort to clarify expectations and needs.

The CAT provides a forum for collaborative learning and a support network for staff as they face data acquisition and policy development challenges. While departments will be providing data, they are also expected to disclose operational information that allows the CAP authors to accurately project emissions and quantify GHG reducing actions already being implemented. The process of disclosing current GHG reducing actions allows departments to learn from each other and bolsters participant confidence in the process. In many cases, these actions are adopted to improve efficiency or to lower costs rather than lowering GHG emissions. Identifying these actions clearly demonstrates to participants that climate action planning is compatible with local operations. For example, the City of Benicia, California, established an interdepartmental team dubbed the Green Team. This team includes representatives from all City departments. One of the activities organized by the team was a speaker series open to all City staff covering climate change prevention-related topics. These presentations served to raise staff awareness of the intention and utility of strategies that reduce GHG emissions. This team was further tasked in 2009 with development of internal emissions-reducing strategies as part of the City's CAP.[6]

While the education process is ongoing there are a few early steps critical to assuring all participants have a shared understanding of the process and the role of CAT. These key phases in the CAT educational process are briefly described here.

Climate Change Science and Policy Overview
Awareness of climate science, emission reducing strategies, and climate adaptation strategies will vary among members of a CAT. A critical first step is assuring a common knowledge base, shared vocabulary, and collective understanding of context. This broad overview can be conducted solely for the CAT or as a series of workshops open to all government staff. These workshops or meetings should cover basic climate science (see appendix A); federal, state, and local policy; and the context in which the jurisdiction sees the plan in relation to other local policy. This staff preparation also lays a foundation for future engagement of the community, particularly the potential presence of climate change skeptics or deniers.

Climate Planning Process
It is critical that all participants understand the overall climate action planning process. Throughout presentation of the process, the role of the CAT should be clearly articulated so that participants are aware of the areas in which they will be contributing. The intention is to provide an overview of the process from project inception to implementation and monitoring. It should provide a clear understanding of the relationship between a GHG emissions inventory, policy audit, and CAP. It should also cover the time horizon expected for implementation and periodic monitoring and reporting. It is here that the link between basic climate science, policy, and emissions estimates can be made tangible for CAT participants. A presentation of the preliminary plan development timeline should also be included at this point to clearly communicate the commitment duration expected of CAT members.

GHG Emissions Inventory
The details of GHG inventory development are covered in chapter 4. During this introductory process, it is critical to communicate the intent and role of the GHG emissions inventory. An overview of the expected data needs can also be covered. Gathering the data for the emissions in-

ventory can be labor-intensive, with data kept in disparate locations. Obtaining information is easiest if the participating departments understand the needed level of detail and the data's intended use. This can be accomplished through a presentation detailing the GHG emissions inventory process, with specific examples demonstrating the use of requested data. Increased knowledge of the process also allows departments to evaluate if more appropriate or additional staff is necessary for CAT membership.

Consider the Logistics of Plan Development

Communities deciding whether to develop a CAP will want to consider factors of cost, time commitment, and needed expertise. GHG emissions inventories require data collection and analysis, specialized software, and personnel with knowledge of government operations and community energy, transportation, and infrastructure data. Development of emissions reduction and climate adaptation strategies may require community education and participation. It will also require specialized knowledge in transportation, energy, building and development, utilities, hazard mitigation, and finance.

Costs of Preparing a Climate Action Plan

Communities will vary in their choice of planning process and type of CAP so costs will vary. Most of the costs associated with preparing a CAP are tied to labor and time. A typical CAP will require 1,000 to 2,500 staff hours. If a consultant prepares the CAP, the cost can range from $50,000 to $300,000. The following sections describe key factors that will affect time and cost.

Level of Public Education and Outreach
Since climate change and climate action planning may be relatively new concepts in some communities, a more extensive public education and outreach effort can be required. In addition, the issue of climate change is a politically charged one that may raise the level and intensity of participation. Many communities have chosen to spend significant up-front time in educating the community and elected officials on the science of climate change and the climate action planning process. This can be one

of the largest time commitments for a CAP and can thus drive costs toward the high end of the range. Chapter 3 provides additional detail on development of a public education and outreach process.

Participation of Advisory Groups

Many communities have chosen to establish formal advisory groups for the CAP. Similar to public education and outreach, the role, membership, number, and schedule of the advisory groups will determine the level of staffing necessary. Some advisory bodies may consist of technical experts who are simply reviewing and commenting on drafts and thus require relatively little staff time. Others may be more policy oriented, may require detailed staff reports, and may hold open debates, thereby demanding significant staff time. In some cases, advisory bodies develop a life of their own and exceed initial estimates of staff time commitments.

Status and Content of the GHG Emissions Inventory

Best practice requires that a CAP be based on a GHG emissions inventory. The level of data collection, management, and analysis required could consume as much as 25% of the time and budget for the typical CAP. In some communities, a community-wide baseline GHG emissions inventory has already been prepared but may need to be updated or peer reviewed to ensure its accuracy and reliability as the foundation for the quantification of strategies included in a CAP. This potential time and cost should be considered as well.

The choices about content and level of detail of the inventory will affect costs. Most CAPs address local government operations emissions and community-wide emissions, thus inventories must be prepared for both, but some only address one of these types. Communities will vary in the quality and accessibility of their emissions data. The GHG emissions forecast can also vary as to whether it is a singular forecast versus a scenario forecast. Chapter 4 provides additional detail on development of a GHG emissions inventory.

Specificity of GHG Emissions Reduction Strategies

GHG emissions reduction strategies in CAPs vary in their degree of specificity. This is partly based on whether the community is preparing a more visionary CAP or one more focused on implementation. Reduction strategies may quantify GHG emissions reductions, compare costs and benefits, and assess feasibility. Each of these would require additional time in preparation. Chapter 5 provides additional detail on development of GHG emissions reduction strategies.

Degree of Contribution by the Climate Action Team

A CAT provides two benefits that will affect the time and cost of the CAP. First, CAT members facilitate a positive and cooperative relationship with their departments. This is necessary because the GHG emissions inventory will require data that may be difficult or time consuming to provide and which departments may not be used to providing. Second, they have the technical depth of knowledge to contribute to development of the GHG emissions inventory and the development of new policies and programs.

Level of Integration into Other Planning Documents

As already discussed, a CAP may be a stand-alone document or it can be integrated with other plans such as a comprehensive or general plan, sustainability plan, or energy plan. Integration will require additional time to ensure consistency including amendment of existing plans that may require additional informal and formal review.

Level of Review Required

In some states, the CAP may require review in addition to the standard local government resolution. For example, CAPs prepared in California are subject to environmental review under the California Environmental Quality Act. In Florida, a CAP that is integrated into a comprehensive plan will require a general plan amendment that is subject to state and regional level review. These additional levels of review will extend the time frame for adopting the plan and will result in additional costs.

Level of Consultant Support

It is increasingly common for communities to hire consultants to assist in preparation of the plan. Some communities hire consultants to prepare the entire plan, while others will only hire for specific tasks. When doing the task mostly in-house, common tasks to hire out include the GHG emissions inventory due to its technical complexity and the public education and outreach program due to the need for specialized expertise.

Table 2.3.   Climate action plan preparation time frame

Phase/tasks	Time to complete
Phase I: Preliminary Activities	3–6 months
1. Establish community commitment	1–2 months
2. Build community partnerships	Ongoing
3. Establish the role of the plan	1–2 months
4. Assemble a climate action team (CAT)	2–3 months
5. Consider the logistics of plan development	1–2 months
6. Establish a public education and outreach program	Ongoing
7. Audit community policies and programs	2–3 months
Phase II: Climate Action Plan Development	9–12 months
8. Conduct a baseline GHG emissions inventory and interim forecast	4-6 months
9. Formulate plan vision and goals	2–3 months
10. Identify a GHG emissions reduction target	1-2 months
11. Develop, evaluate, and specify GHG emissions reduction strategies	4–6 months
12. Quantify GHG emissions reduction strategies	2–3 months
Phase III: Implementation and Monitoring	Ongoing
13. Develop and administer an implementation program	Ongoing
14. Monitor and evaluate implementation and reduction target attainment	Every 1–2 years
15. Modify and update the plan	Every 2–5 years

Time Needed for CAP Preparation

Based on the foregoing process steps and factors, the CAP will usually take one to one and a half years to prepare and adopt. Table 2.3 shows estimates for each of the process steps. Since some tasks can be completed concurrently, the phase I timeline is shorter than the sum of the individual tasks (fig. 2.1).

Funding Options for CAP Preparation

A critical issue for development of the CAP is how it will be funded. Although some communities have enlisted significant volunteer support for the effort, particularly from colleges and universities, there will likely

Figure 2.1   Example of phase I timeline.

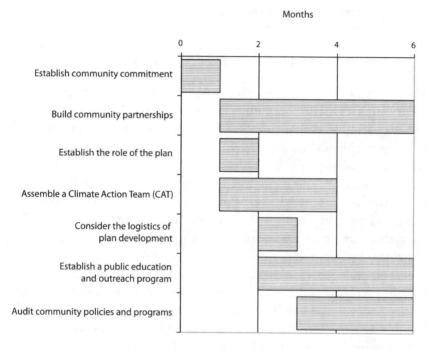

be some costs incurred. But for most communities the costs discussed at the top of this section will have a significant impact on the scope and quality of the climate action planning effort. There are several possibilities for funding, including local government general funds, fees, private foundation donations and grants, and state and federal grant programs, but these change often. Recently many communities across the nation used federal Energy Efficiency and Conservation Block Grants (EECBGs) funds for preparation of their CAPs. These funds were authorized as part of the American Recovery and Reinvestment Act (Recovery Act) of 2009 and either went directly to communities or were channeled through state programs. The EECBG funds assisted communities to address energy efficiency, GHG emissions reduction, and job creation; this included the ability to fund preparation of a CAP. The program was funded at $3.2 billion, and allocations were made to eligible governmental agencies, including cities and counties, through formulas and through competitive grants. Its future availability is uncertain and will be decided by the president and Congress. Communities are best advised to explore all opportunities and talk to other communities about how they funded their CAPs.

Audit Existing Community Policies and Programs

The policy audit assesses preexisting policies, programs, and procedures for consistency with community goals for GHG emissions reduction and adaptation to climate impacts. For example, many communities already have policies and programs aimed at promoting transit use, bicycling, and walking—the type of goals that may be found in a CAP. By going through the audit process, a community can establish the local policy context in which the CAP must be devised, adopted, and implemented. The policy audit is also a chance for a community to describe the great things it is already doing to address GHG emissions reduction and climate adaptation.

The audit provides useful information for several climate action planning activities. First, many communities will have implemented activities between the baseline emissions year identified in the inventory and the current year. The policy audit provides the information necessary to estimate emissions reductions achieved in that time and forecast the long-term reductions likely to result from those activities. Second, this information can help a community set a more realistic emissions reduction target that accounts for existing and proposed polices that reduce GHG emissions. Third, the policy audit lets the community clearly identify the gaps in its current policy and program framework. Thus the CAP can focus on filling these gaps.

Elements of a Policy Audit

The policy audit typically assesses the local government's policy and operational procedures that may or may not be formally documented. A thorough policy audit requires close collaboration with the CAT whose members aid in the identification of relevant policy documents and disclose operational procedures. In addition, though not yet common, a community may wish to account for nongovernmental programs that reduce GHG emissions such as CFL giveaway programs from electricity providers or energy conservation education programs offered by nonprofits. The audit seeks to identify policy or programs that already serve to reduce GHG emissions and identify policy or programs incompatible with the goals of addressing emissions reduction and climate adaptation. It is best to organize the audit into sections that generally reflect those in the emissions inventory (i.e., transportation, water, energy, etc.). This division helps in confining assessment to policy that directly impacts emission sources. The following sections explain the local government policies and operations audit.

Adopted Policy
Adopted policy refers to plans, ordinances, and other local government laws that directly influence GHG emissions. The content of these policy documents falls into two broad categories: supporting policies that act to reduce GHG emissions and potentially conflicting policies that either directly increase emissions or prevent emissions–reducing actions (box 2.2). Supporting policy identifies elements that do not need to be included in the CAP or areas to be enhanced. Potentially conflicting policies identify those areas where plans or other city policy will require amendment to become compatible with CAP goals. It is critical that the evaluation be confined to actions that *directly* impact GHG emissions or sources of GHG emissions. It is quite easy, through a series of hypothetical scenarios, to tie any local government policy to GHG emissions. The inclusion of far-fetched causal links makes the audit convoluted and more difficult to use as a starting point for CAP development. It is critical that each policy be tied to items in the GHG emissions inventory. Table 2.4 lists several of the documents most likely to be important components of an audit of adopted policy.

Identifying current policy is the first step in conducting the audit. Existing policy that supports GHG mitigation must be further divided based on level of implementation. Policy that has been or is currently being implemented should be separated from adopted policy that has not been funded or implemented. Implementation can be assessed through examination of funding allocation in the local government budget, periodic reports on implementation of plans, and feedback from the CAT. The CAT is a critical check in assuring that this division is done accurately.

Operational Procedures
Many of the actions taken by local agencies that reduce GHG emissions are enacted not only for environmental reasons but also in the interest

Box 2.2
Example of Supporting and Potentially Conflicting Policy

Policies, such as those that encourage energy efficiency and promote walking, biking, or public transportation over automobile trips, will directly reduce greenhouse gas emissions. In a few cases, a city's policies and programs support actions that may generate additional emissions, such as those that encourage vehicle travel.

Supporting: General Plan Policy X.X: No urban development beyond the urban growth boundary shall be served by city water and/or sewer services.

This policy is one that assures dense development, curbing commute distance. Because this action directly influences resident vehicle miles traveled, it should be included in the audit as a supporting policy.

Potentially Conflicting: General Plan Policy X.X: Maintain at least Level of Service D on all city roads, street segments, and intersections.

The most common way to improve roadway Level of Service is to increase capacity often with increased number of lanes. While such actions may increase vehicle speeds in the short term, which reduces emissions, in the long term it can increase vehicular traffic.

of efficiency and cost savings. This information can only be obtained through close interaction with the CAT. Many of the procedures may be standard protocol for individual departments. This will include the maintenance schedule for municipal structures, which may entail actions such as replacement of heating, ventilating, and air conditioning systems or the addition of motion sensors for lighting. It will also include the vehicle fleet turnover schedule and purchasing policy. In many cases, saving money on energy to operate buildings and fuel to power vehicles is motivating these choices, and staff may not initially recognize the benefit in terms of GHG emissions. Another area to evaluate is employee programs such as transit incentives, office education programs, and environmentally friendly purchasing programs.

Many of the operational procedures will be revealed during development of the emissions inventory, but some of these programs may have been created in the time since the baseline year. In other cases, they may not be actions that directly emit GHGs. For example, an emissions

Table 2.4. Summary of some of the documents most commonly included in a policy audit

---

Plans

Comprehensive Land Use Plan (General Plan) Parks, Trails, and Open Space Master Plan Urban Water Management Plan Bicycle and Pedestrian Master Plan Transit Plan Area Plan Downtown Mixed-Use Plan Local Hazard Mitigation Plan	Plans are necessarily broad; therefore, focus should be placed on the lower levels of policy hierarchy (policies and programs, as opposed to goals or objectives). The goal is to identify plan elements that directly influence emissions reduction or increase and community risk or resiliency from climate change.

Standards, ordinances, programs, and policy

Zoning Code Green Building Ordinance Water-Efficient Landscape Standards Tree Ordinance Environmentally Friendly Purchasing Policy Traffic Calming Program Floodplain Ordinance	These are the regulations that implement the plans. In these documents one may find the specifics lacking in the plans. The specificity of these regulations allows the assessment to be more nuanced.

Memoranda, feasibility assessments, and other nonbinding statements

---

Community Garden Memorandum Energy Efficiency Status Memorandum Feasibility Assessments	While these documents are not enforceable policy, they do provide a good indicator of city perspective and awareness. For example, a city that has conducted a feasibility study on local renewable energy generation may not have installed any renewable energy projects, but the study may signal a political willingness to do so.

---

inventory will include energy consumed by municipal buildings. But whether the energy use data are higher or lower than normal due to an education campaign or installation of motion detectors for lighting is irrelevant to the accuracy of the inventory. When a CAP is being developed, these data are critical for choosing strategies to reduce energy use.

From Policy Audit to CAP Development

The policy audit, containing a review of adopted policy, guidance documents, and department operations, should be submitted to the CAT as well as the departmental staff. This will be a final check on accuracy by those tasked with implementing the reviewed plans, guidance, and protocols. This also serves to assure that members of the CAT share a common basis for CAP development, review, and implementation. The policy audit should set the stage for CAP development by identifying areas in which short-term strategies can be most effectively pursued and also where there are gaps in or conflicts with current policies. The policy audit data inform the prioritizing of strategies in the CAP.

Next Steps

Each community will need to tailor these preliminary activities to its own needs and interests. It is good practice to talk to similar communities that have gone through the climate action planning process about how they got started and what they would do differently in hindsight. Communities that work through the issues raised in this chapter will find that the rest of the planning process will go more smoothly and that the final CAP will be a better product. The next several chapters describe phase II of the planning process: CAP Development.

Chapter Resources

Climate Action Planning Guides

ICLEI–Local Governments for Sustainability, *U.S. Mayors' Climate Protection Agreement: Climate Action Handbook* (Oakland, CA: Author, n.d.). http://www.iclei.org/documents/USA/documents/CCP/Climate_Action_Handbook-0906.pdf. This handbook provides a basic and accessible introduction to climate action planning and GHG emissions reduction strategies.

American Planning Association, *Policy Guide on Planning and Climate Change* (Washington, DC: Author, 2008). http://www.planning.org/policy/guides/pdf/climatechange.pdf. This guide makes an argument for why city planners should be involved in the climate action planning process. It also provides an extensive policy framework for developing GHG emissions reduction and climate adaptation strategies.

Walter Simpson, *Cool Campus! A How-To Guide for College and University Climate Action Planning* (Lexington, KY: Association for the Advancement of Sustainability in Higher Education, n.d.). http://www.aashe.org /files/resources/cool-campus-climate-planning-guide.pdf. This guide is specifically aimed at assisting colleges and universities that want to prepare CAPs.

Pew Center on Global Climate Change, "Climate Change 101: Understanding and Responding to Global Climate Change" (series). http://www .pewclimate.org/global-warming-basics/climate_change_101.

Background on Sustainability and Urban Planning

Philip R. Berke and David R Godschalk, *Urban Land Use Planning*, 5th ed. (Chicago: University of Illinois Press, 2006).

Urban Land Use Planning, often referred to as the planner's bible, is a comprehensive source of the planning process, translating theory into methods and actions. The fifth edition includes new land use planning issues for the twenty-first century, including how to incorporate the "three Es" of sustainable development (economy, environment, and equity) into sustainable communities, methods for including livability objectives and techniques, and the integration of transportation and land use.

Tommy Linstroth and Ryan Bell, *Local Action: The New Paradigm in Climate Change Policy* (Burlington: University of Vermont Press, 2007). *Local Action* provides an overview of climate change science and policy at the local and federal levels, presents various emissions reduction strategies for municipalities, and offers case studies from the cities of Fort Collins, Colorado, and Portland, Oregon.

Peter Newman, Timothy Beatley, and Heather Boyer, *Resilient Cities: Responding to Peak Oil and Climate Change* (Washington, DC: Island Press, 2009). *Resilient Cities* presents climate change and peak oil as a double whammy that will force cities to become more resilient or face the potential for collapse. It creates a potent argument for the need for change and then lays out a vision for that change including "10 strategic steps."

Stephen Wheeler, *Planning for Sustainability: Creating Livable, Equitable and Ecological Communities* (New York: Routledge, 2004). *Planning for Sustainability* defines and makes the argument for sustainability as a key component of city planning. It is primarily a resource book for ideas on implementing sustainability in planning at several different levels of governance.

Chapter 3

Chapter 3

Public Participation

Climate action planning should include public participation. The United States is a democracy in which the public has a right to participate in the activities and decisions of the government. On a more practical level, many aspects of CAP implementation require community members to voluntarily change behavior in areas such as choice of transportation mode and indoor energy usage, and local organizations are needed to support these changes. As a result, successful implementation of GHG reduction strategies in a community will rely on direct engagement with the public and other community entities throughout the planning process. Public participation can result in a better plan, legitimize the plan in the eyes of the public, gain "buy-in" from the public, and ultimately build social capital in the community. Public participation has become standard practice in the preparation of CAPs. In many communities, public task forces are the main drivers behind the preparation and adoption of a CAP.

Because participation is regularly included in all types of plan development, the field of planning has a well-developed theoretical understanding of the role, assumptions, and characteristics of varying depths of participation. Best practice calls for establishing opportunities for the public to participate in the planning process. In some communities, however, there may be a reluctance to include the public given concerns that an uninformed public could slow down or derail the planning process, or a perception that the public isn't interested in participating.

As to the first concern, it is true that a quality public participation program will often take more time; it is also true that it can improve implementation and ultimate effectiveness of the plan.

As to the second concern regarding the level of public interest, there is good evidence that the public and other community entities (e.g., small businesses) are interested in participating in deciding the future of their communities.[1] They are not apathetic or unconcerned; they simply want assurance and ongoing reinforcement that their input is meaningful. Most people have very busy lives, and taking a night off to attend a meeting about climate action planning may not be a priority, especially if they don't believe their concerns will be considered.

Talking about Climate Change with the Public

Communicating with the public about climate change can be more challenging than many other planning endeavors. Climate change generally has a lower level of public interest than other public policy issues. According to several recent polls, public opinion regarding climate change is divided and fluid.[2] A slight majority of Americans tend to believe that global warming is occurring, although uncertainty is very high, especially when people are asked whether it is being caused by human actions. Since 2007 there has been a clear trend toward greater skepticism regarding the occurrence and importance of the global warming phenomenon. One of the first tasks of the climate action planning team is to know the views of their intended audience. If there is not a good sense of the public's opinion, then a community survey may be useful.

Research has identified the range of views community members may have regarding climate change. A report from Yale and George Mason Universities provides national averages for how a community may be differentiated based on members' beliefs. The report identifies "Six Americas":

> The Alarmed (18%) are fully convinced of the reality and seriousness of climate change and are already taking individual, consumer, and political action to address it. The Concerned (33%)—the largest of the six Americas—are also convinced that global warming is happening and a serious problem, but have not yet engaged the issue personally. Three other Americas—the Cautious (19%), the Disengaged (12%) and the Doubtful (11%)—represent

different stages of understanding and acceptance of the problem, and none are actively involved. The final America—the Dismissive (7%)—are very sure it is not happening and are actively involved as opponents of a national effort to reduce greenhouse gas emissions.[3]

While the distribution of these views will vary by community, all communities will likely contain some portion of the public and decision makers in each group.

The planning team must be prepared to communicate to each of these groups. The Alarmed and Engaged will need little information on the science of climate change and will instead mostly focus on any information that lets them better understand options for reducing greenhouse gas (GHG) emissions. The Cautious and Disengaged, if they can be reached, may need information focused on the impacts of climate change to them and their communities. The Doubtful and Dismissive may need information on the basic science and evidence of climate change. A clear message for climate action planners that comes from the surveys is that considerable uncertainty exists, and many Americans feel they need more information to better understand and judge the issue.

Or not. It is also argued that stagnation in the CAP development process driven by differences over the scientific facts of climate change is not the problem.[4] Instead the problem is poor policy design. In other words, many strategies that can reduce emissions or community vulnerability to climate impacts also have many other benefits. Perhaps, for community members that could be categorized as Doubtful or Dismissive the best strategy is to point out and focus on the range of co-benefits that result from CAP development rather than try to convince them of the science. In some communities, a CAP is given a different title that focuses on energy such as the Newton, Massachusetts, *Energy Action Plan*. It may be possible that community members with divergent views of climate change may reach consensus on the need for action in some sectors if the co-benefits are made clear. It is the job of the participation facilitator to aid community members in identifying such common ground.

Americans tend to view the global warming problem as long term rather than imminent; thus they tend to deprioritize the issue. There is also a skepticism that anything can be done about it even if global warming is a problem. About half of Americans support policy to reduce carbon emissions, although this support is tempered if it directly costs them. There is some evidence that opinion on these issues varies by political orientation and age but not level of education. The older and more politically conservative tend to be more skeptical of climate change. The surveys also indicate a public belief that there is considerable disagreement among scientists, but recent surveys of scientists working in climatology show a strong consensus that the planet is warming and that human activity is a significant factor.[5,6] Since most people generally trust scientists as a source of information about climate change, getting scientists involved may be an important part of a public education program.

To address these concerns, it is important to establish three content goals for communicating climate change science and climate action planning to the public:

- Clarify consensus on, and uncertainties in, the science.
- Explain why the problem is important, especially to them personally.
- Explain that the problem is solvable.

Each of these content goals relies on the planning team's ability to effectively communicate science to citizens. In preparing materials and activities to present science in a manner understandable to participants, it is important to be aware of the criteria that a public audience uses to make judgments about science. These criteria include the following questions:[7]

- Does scientific knowledge work? Do public predictions by scientists fail or prove to be true?
- Do scientists pay attention to other available knowledge when making claims?
- Are scientists open to criticism? Are they willing to admit errors and oversights?
- What are the social and institutional affiliations of scientists? In other words, do they have a historical track record of trustworthiness? Similarly, do they have perceived conflicts of interest relative to their associations with industry, government, universities, or advocacy groups?

- What other issues overlap or connect to a public's immediate perception of the scientific issue?
- Specific to risks, have potential long-term and irreversible consequences of science been seriously evaluated, and by whom? And do regulatory authorities have sufficient powers to effectively regulate organizations and companies who wish to develop the science? Who will be held responsible in cases of unforeseen harm?

Approaches to Public Participation

A review of recently completed CAPs across the United States reveals three general approaches to public participation:

- Green Ribbon Task Force
- Public Task Force (working groups)
- Community Workshop

In the Green Ribbon Task Force approach, a select group of community leaders and specialists representing such areas as government, business, industry, science, agriculture, schools and higher education, and environmental and community groups are assembled to guide development of the plan. Members of the Green Ribbon Task Force are usually selected or appointed by elected officials. In a study by the authors, the size of these task forces ranged from twelve to thirty-six people. Although there may be other opportunities for the public to be involved in the planning process, the task force provides the primary method of participation for the diverse interests in the community. The planning team serves as staff for the task force and makes major decisions affecting the development of the CAP. The sectors represented by the task force are viewed as critical collaborators for effective plan implementation. This approach was exemplified in communities such as Chattanooga, Tennessee (box 3.1), Denver, Colorado, and Tacoma, Washington. One caveat here is that if a Green Ribbon Task Force is small and not representative of the interests of most of the community it may not achieve these goals. The Green Ribbon Task Force may need to create opportunities for input from more diverse voices from the community.

The Public Task Force is more inclusive and usually much larger than the Green Ribbon Task Force. The participants include the types

Box 3.1
Example of the Green Ribbon Task Force Approach

The Chattanooga (Tennessee) Climate Action Plan

In 2007, Mayor Ron Littlefield appointed the Chattanooga Green Committee to advise him on which climate actions the community should take. The Committee met for one year to analyze greenhouse gas emissions and other community data, collaborate with the community, meet with subject matter experts, and develop a set of recommendations. The Committee included fourteen members representing private business, the public sector, and educational institutions. They divided themselves into four task forces: Energy Efficiency, Healthy Communities, Natural Resources, and Education and Policy. The Committee was composed of representatives from the following groups:

- City of Chattanooga, Division of Urban Forestry
- Advanced Transportation Technology Institute, Chattanooga Technology Council
- U.S. Green Building Council
- Hamilton County Government
- Volkswagen Group of America, Inc.
- Chattanooga Manufacturers Association
- Chattanooga-Hamilton County Air Pollution Control Bureau
- Chattanooga Tree Commission
- Chattanooga Home Builders Association
- Chattanooga City Council
- Associated General Contractors of East Tennessee
- University of Tennessee at Chattanooga, Department of Biological and Environmental Sciences
- Electric Power Board

of stakeholders already listed, as well as ordinary citizens. The Public Task Force membership is often self-selected or voluntary rather than appointed by elected officials, though elected officials may confirm or authorize creation of the Public Task Force. The assembled task force does much of the work of developing and vetting climate policies and actions and may be broken into subcommittees by sector or strategy area. The task force may also report to an organizing or steering committee (or even a Green Ribbon Task Force). The planning team provides support for the task forces and depends on the task forces to do the heavy lifting on plan development. This approach was exemplified in Cincinnati, Ohio. In 2007, the mayor of Cincinnati, Mark Mallory, appointed the Climate Protection Steering Committee to assist in developing the plan. The Committee was composed of business, government, environmental, academic, and civic organization leaders. Although this is similar to the Green Ribbon Task Force Approach, the Committee formed five Task Teams in the areas of Energy, Transportation, Land Use, Waste Management, and Advocacy and empowered 150 subject matter experts and concerned citizens to compile hundreds of potential actions for the city.

In the Community Workshop Approach, public participation is primarily accomplished through a series of community workshops or forums where anyone may participate. Meeting information may be supplemented with online forums and surveys. The planning team has primary responsibility for plan development and uses the meetings to share information and ideas, assess preferences from the community, and receive feedback on proposed strategies. This approach was exemplified in communities such as Hayward, California, and Madison, Wisconsin.

In some communities public participation is minimal and may only be done to meet the legal requirements of the adopting organization. If the plan is being prepared or adopted by a governmental agency, there are likely state or local laws with specific, minimum requirements for public participation. Most states have laws that require a government agency to notify the public of pending actions (usually with a legal notice in the newspaper at minimum) and to hold open public hearings where the public may provide testimony. This is usually seen as an absolute minimum for public participation in climate action planning. Best practice is to exceed this minimum with more inclusive and meaningful forms of participation.

These do not constitute all the possible approaches for public participation. There are many ways to engage the public and many communities choose to combine or modify elements of the three common approaches.

The following section examines five critical choices in designing participation programs and explains their relevance to each of the three recommended approaches.

Critical Choices in Designing a Public
Participation Program

A series of critical choices must be made in designing participation programs:[8]

1. *Administration*: whether or not to prepare a participation program and how to staff involvement efforts
2. *Objectives and purpose*: what the participants are being asked to do and whether to educate, seek their preferences, and/or grant them direct influence
3. *Stage*: when to start encouraging citizen participation in the planning process
4. *Targeting*: which types of stakeholder groups to include in participation efforts
5. *Techniques*: what types of participation approaches to employ and what types of information and dissemination processes to incorporate into participation activities

These choices should be made based on a review of practices in the field and adherence to the "characteristics of highly effective public participation programs" established by the Department of Energy:[9]

•   Have a clearly defined expectation for what they hope to accomplish with the public
•   Are well integrated into the decision-making process
•   Are targeted at those segments of the public most likely to see themselves as impacted by the decision (stakeholders)
•   Involve interested stakeholders in every step of decision making, not just the final stage
•   Provide alternative levels of participation based upon the public's level of interest and reflecting the diversity of those participating
•   Provide genuine opportunities to influence the decision
•   Take into account the participation of internal stakeholders as well as external stakeholders

These criteria can be used to design and evaluate a public participation process. Appendix B provides a model approach for a public participation program that can be tailored based on the public participation approach desired and addresses these program choices and criteria.

Administration

Assuming a community has chosen to have a public participation process for developing the CAP, the first issues to address are staffing and budget. Staff identification and budget needs are based on the desired scope of the participation program and should be part of the initial CAP timeline and overall budget. The staff and allocation of funds should allow for flexibility in the participation process. Issues or stakeholder groups not identified during the initial planning phase may emerge. In this case, the staff and funds allocated for participation should be able to accommodate the inclusion of additional information or events (e.g., meetings, workshops, etc.).

Equally important as having adequate staff hours and funding is choosing the person or organization that will oversee, convene, facilitate, and record outcomes. The participation process can be led by the planning entity that is overseeing CAP development (e.g., a city, county, or consultant), or an outside entity can be brought in. This outside entity can be a local nonprofit organization or a consultant specializing in community participation. Use of a local entity can either build or threaten community trust depending on the manner in which the organization is viewed, particularly by the stakeholder groups targeted by the participation process. An outside consultant may benefit from being seen as neutral, but a lack of familiarity may result in reluctant participation by stakeholders.

Participation events bring together a collection of stakeholders with potentially divergent views. The organizing entity must have the facilitation skills to ensure that all participants feel involved and heard, particularly for participation efforts that not only seek to inform but also solicit feedback. The views expressed must be recorded so that participants can evaluate whether or not they have been accurately heard, and the feedback can be referenced in the future. In addition to facilitation skills, the individual or group that leads participation events in support of CAP development should have a clear understanding of climate science, the CAP development process, and the local role of the CAP. This is important as community concerns and desires for the future

Figure 3.1 Range of public participation options.

	Inform	Consult	Involve	Collaborate	Empower
Participation Goal	To provide information on climate change, CAP development, and local options.	To gain feedback on local goals and proposed strategies.	To be involved throughout the CAP process to assure community concerns are understood and considered.	To facilitate community collaboration in each aspect of CAP devleopment including goals & strategy development.	To grant final decision making to the community.
Commitment to Participants	Will be kept informed.	Will be kept informed with local concerns & aspirations acknowledged & opportunities for feedback provided.	Will work to ensure that local concerns & aspirations are directly reflected in the strategies developed and the role of input in the decisions is clearly communicated.	Will seek consensus on identificaion of local needs, planning process, and CAP strategies.	Will decide the strategies to be implemented.
Example	Fact Sheets Websites Reports Press Release Mailers Exhibitions/Speakers Citizen Academy	Exhibitions Public Hearing Workshop Survey Comment Boards	Visioning Workshop Focus Groups Survey	Visioning Workshop Focus Groups	Visioning Workshop Task Forces

Source: Modified from "Spectrum of Public Participation," International Association for Public Participation, accessed August 23, 2010, http://www.iap2.org/displaycommon.cfm?an=5.

may not appear directly related to GHG emissions or climate impacts. A good facilitator can aid in public understanding by explaining these connections. The oversight and facilitation of these events need not be undertaken by a single individual. A team approach may best provide the skills necessary to meet participation goals.

Objectives and Purpose

The initial tasks of designing the participation program are to identify the desired outcomes, define the depth of participation, and establish the level of commitment being made to the public. The range of participation options varies based on the manner in which a community addresses these considerations (fig. 3.1).[10] The depth of participation is determined by (1) the extent of the opportunities offered to those who chose to participate, and (2) the weight given to the views expressed by participants.[11,12] Another way to conceptualize depth of participation was established in Sherry Arnstein's famous "Ladder of Citizen Participation."

The ladder has eight rungs for each level of public participation and these are grouped into three categories: citizen power, tokenism, and nonparticipation. The level of public influence or control in the planning process is highest towards the top of the ladder and decreases toward the bottom:

Citizen power
Citizen control
Delegated power
Partnership
Tokenism
Placation
Consultation
Informing
Nonparticipation
Therapy
Manipulation

Arnstein proposes that for any planning process the following question could be asked: Where does the power to control the planning process reside? Or alternatively: At what level (or to what depth) is the public participating in the planning process? Based on the ladder analogy, she established that citizens who were being placated or manipulated through participation schemes were at the bottom of the ladder, whereas citizens who were given full partnership and power in planning and decision making were at the top of the ladder. She suggested moving up the ladder as much as possible when involving the public in community issues. Each community has to decide its own desired level of participation, then clearly communicate to members of the public their role in the process and the commitment being made regarding their role and the use of their input.

Depending on the goal of a participation process, a community can develop a program using one or more of the three approaches to participation (Green Ribbon Task Force, Public Task Force, or Community Workshop) and select complementary levels of opportunity and event types. For example, if the goal is to inform, then a basic information distribution campaign through mailers or press releases may be sufficient. If the goal is to consult or involve, then the Community Workshop Approach may produce the best results. If the goal is to collaborate or empower, then either the Public or Green Ribbon Task Force Approach may create the most effective partnership. Communities with multiple goals for their public participation programs may find that blending one or more of the approaches would provide a robust and comprehensive approach.

Stage

Deciding when to include the public in the climate action planning process will be partially determined by the objectives discussed in the previous section; the greater the depth of participation, the greater the need to include the public early and often. Early on there may be a need for a kickoff meeting. Once the planning is under way there may be a need for periodic meetings or events. Once the draft plan is completed there may be a need for final meetings leading up to adoption by the relevant organizations or local governments. Each of these three stages—kickoff, planning, adoption—should be considered for public participation.

During the kickoff phase, public participation will usually focus on education and outreach. Since many in the public are unfamiliar with climate action planning, the participation process should begin by educating stakeholders about climate change and climate action planning. Moreover, participants should understand why climate action planning is needed and why it is relevant to their lives, with a particular focus on co-benefits. In addition to education and outreach, it may be necessary to convene the public to actively engage in a visioning process (box 3.2). Visioning is the collective exercise of describing the desired outcome or a desired future that the CAP could help achieve. For example, a vision statement might include the idea that the community prioritizes transportation systems for bicyclists and pedestrians. Finally, the kickoff meeting can be used as an opportunity to recruit members of the public to serve on public task forces as was done in Evanston, Illinois.

The planning phase will usually focus on having the public participate in the development of ideas for GHG reduction and adaptation

Box 3.2
City of Berkeley (California) Climate Action Plan
Vision for Year 2050

- New and existing Berkeley buildings achieve zero net energy consumption through increased energy efficiency and a shift to renewable energy sources such as solar and wind.
- Public transit, walking, cycling, and other sustainable mobility modes are the primary means of transportation for Berkeley residents and visitors.
- Personal vehicles run on electricity produced from renewable sources or other low-carbon fuels.
- Zero waste is sent to landfills.
- The majority of food consumed in Berkeley is produced locally—that is, within a few hundred miles.
- Our community is resilient and prepared for the impacts of global warming.
- The social and economic benefits of the climate protection effort are shared across the community.

strategies that could be in the CAP. This phase can vary in length and is often iterative. Brainstorming and other idea-generating efforts would be facilitated during this phase, including ongoing education about issues and updates on the CAP development progress. The ideas generated though the brainstorming process can be developed into preliminary strategies. These strategies can be presented back to the community to assure that the ideas expressed in the brainstorming and other events were accurately understood. This is also a good time to identify any gaps that the community sees in the potential suite of strategies.

The adoption phase occurs when the CAP strategies are nearing completion. This stage usually garners the most attention from the public and thus may be the point of most controversy. A greater depth of participation in the earlier stages can serve to limit the level of controversy in this stage. In the adoption stage a draft CAP is available, and the public reacts to the information that it contains. Participation efforts will again focus on education and outreach. In addition, the entity that is leading the participation efforts should be prepared to facilitate and mediate meetings and disputes about the contents and direction of the CAP.

Targeting

In the targeting stage, it is important to ask: Who is the public that is involved? Are they representative of the elite, such as community leaders in business and the nonprofit sector with access to considerable resources, or, are they representative of the general public?

In the Green Ribbon Task Force approach it is most challenging to ensure that all community interests are represented. Establishing who is on—and not on—the task force could become a contentious and politicized issue.

In the other two approaches, Public Task Force and Community Workshop, where meetings are generally open to all comers, this issue may be seen as less of a concern, but this would be a mistake. A well-known phenomenon in local government is the lack of diversity at open public meetings. Meetings can be dominated by well-organized community groups and skew significantly on key demographics such as age, income, and housing tenure (renter vs. owner) (table 3.1). Achieving diverse community participation requires good outreach efforts and surveying (formally or informally) of key demographics at meetings to check on representativeness. If a community has members with first languages other than English, then consider having translated materials and simultaneous translation available at meetings. In addition, diversity in meeting structure, location, and date and time can engage a greater portion of the community.

Techniques and Information

The question of which participation approaches to employ is challenging due to the wide variety of choices. This section presents a selection of techniques split loosely into two categories (table 3.2): (1) basic techniques that should be employed in all public participation programs, and (2) advanced techniques that can be useful in some public participation programs.

Basic Techniques That Should Be Employed in All Public Participation Programs
Early in the participation process, the community needs to be notified that CAP development is being pursued. Only after community members are aware of the problem being addressed and the local strategy being employed will they feel compelled to participate in the process.

Table 3.1   Issues affecting public participation rates

Demographic	Issues
Age	Retirees may be more likely to participate during the day or early evening. Working age adults and students may not be able to participate during the day.
Family status	People with younger children may not participate due to time needs of the children.
Income	People working two jobs or unable to afford child care may not participate. People may not have access to the Internet.
Race/ethnicity/culture	People in the minority or who have been traditionally disenfranchised may be skeptical of participating. Some cultures are uncomfortable or unfamiliar with participation in community affairs.
Language	People who have English as a second language may not know about meetings or may believe they cannot participate, especially if translation services are not available.
Housing tenure	People who rent may miss notices mailed to property owners or utility payers. They may also feel they are outsiders.
Residency	New members of the community may not yet be engaged with community planning efforts.

The goal of the basic techniques is to communicate climate science, anticipated local impacts, and the climate action planning process. The intended outcome should be increased community understanding of the issue, local needs, and the local planning process to address the needs. This is also the best time to publicize the participation events and anticipated role of the participants.

Information documentation includes the provision of hard copy and web-based plans, reports, and the like. Hard copies of planning documents should be available at city hall and public libraries at a minimum. The creation of a web page can support this process by serving as a repository for all information relating to the CAP. The site should include presentations, materials, and press releases produced for the meetings as well as versions of the CAP document for public review. The site

Table 3.2   Techniques for public participation

Basic techniques	Advanced techniques
Information documentation	Visioning
Media and public relations	Focus groups
Educational meetings and workshops	Questionnaires
Exhibitions and events	Computer-based polling
Public hearings	Social marketing/Web 2.0
Mail and e-mail notices	Citizen academy
	Competitions and challenges
	Speakers bureau
	Roving workshops

and all materials should have a consistent "look and feel" to them that defines them as being materials representing the CAP project. A website should be accompanied by distribution methods that effectively reach community members who do not have easy Internet access. Materials should be created in different languages for community members who do not speak English.

Media and public relations campaigns engage the media and inform the public through newspaper ads, TV ads, press releases, mailers, community calendars, op-eds or letters to the editor, and websites.

Educational meetings and workshops provide opportunities for members of the public to learn in depth about climate change and CAP development and engage in dialogue with experts and fellow citizens. These can range from large, well-publicized meetings intended to attract a cross-section of the community to smaller meetings for specific groups or neighborhoods.

Exhibitions provide opportunities to display information in public areas, including informational booths at farmers' markets or other regular community events, explaining the plans, processes, and locally important issues. These displays can also be used to gauge public opinion and gather feedback on potential strategies. These opinions can be gathered by having a staff member on hand to answer questions and solicit feedback or through use of a comment board or other self-service means of expressing views.

Public hearings at meetings of elected and appointed boards can also be used as a forum to solicit formal feedback from the public in a manner that becomes part of the official local government record.

Mail and e-mail notices can raise awareness and inform community members on the CAP development process as it progresses. Regular e-mail notices are a simple and low-cost tool to engage community members who are unable to regularly attend participation events such as workshops. E-mail addresses of interested stakeholders can be gathered through other outreach tools such as websites, workshops, and exhibitions.

Advanced Techniques That Can Be Useful in Some Public Participation Programs
Some participation programs seek not only to educate the community and raise awareness but also to solicit more detailed feedback on various aspects of CAP development. The feedback can occur at all stages of CAP development from visioning for the community to individual concerns or specific strategies. These techniques often work best when accompanied by many of the basic techniques just described.

Visioning is a public meeting process that asks citizens to create a vision for a desired future state of their community. Visioning exercises require the public to think less about immediate problems and constraints and instead imagine how they would like their community to be in 20 to 50 years. A vision then becomes a long-range goal or aspiration that a community can move to through deliberate action. The outcomes from a visioning meeting can help focus CAP strategy development and identify key co-benefits.

Focus groups are a research method for gauging public interests and opinions. A community using this approach could conduct a series of stakeholder discussions with key target audiences (e.g., agricultural community, local builders, etc.) with the goal of gathering detailed and specific feedback on the CAP development. Focus groups can also be used at general stakeholder meetings where community members can divide according to interest, expertise, or areas of concern. Stakeholder discussions can focus on proposed methods to reduce GHG emissions from each of the primary CAP sectors. These discussions should be facilitated to encourage substantive discussions of policy implications and benefits within each sector, with the goal of achieving consensus on policy direction for each sector addressed by the CAP.

Questionnaires or surveys can be conducted through a website, mailed in hard copy, or delivered in person. This tool can be used to assess community habits and values, as well as gather feedback on the CAP sector issues. The intended use of the survey results will determine the need for a statistically valid survey or more qualitative, informational survey.

Computer-based polling is a type of questionnaire or survey that can be administered real time during a meeting to provide instant feedback from an audience. Meeting participants can be given small, remote devices (they look like small television remotes) that they use to respond to questions. The computer collects the public responses and immediately calculates and displays summaries of the responses. Real-time polling is also anonymous, which often neutralizes extreme positions and provides equal participation to all participants.

Social marketing/Web 2.0 refers to the use of Internet-based social networking and digital communication tools such as blogs, wikis, collaborative software, and social networking sites such as Facebook and Twitter, and other similar tools. These digital media can be used to keep the public informed, facilitate involvement, or solicit input, especially from younger members of the community who may be less likely to participate in other ways. For example, a Facebook page can be created for a CAP. Developing and posting messages on these sites can encourage conversation on the issues as well as promote the public meetings. Key community-based organizations with a vested interest in the issues discussed in the CAP can be "friended," "liked," or "followed" so as to promote the meetings and other participation events.

Citizen academy is a training course provided for community members that allows for a greater level of detail and depth than can be provided through workshops. For climate action planning, a science academy could address climate science, emissions reduction and climate adaptation strategies, local government policy and planning, and "green" living. Experts could be brought in to teach different parts of the course; local colleges and universities are often an excellent resource.

Competitions and challenges can take many forms. Friendly competition or prizes can help encourage attendance at participation events. Games can provide an engaging and fun context in which community members can be introduced to new concepts or examine the pros/cons of proposed strategies.

Speakers bureaus, organized in advance of community workshops or town hall events, can enable the planning team to connect with key stakeholder groups on important issues and help to inform citizens about the local issues related to climate change and land use planning. A speakers bureau may consist of staff, representatives of key stakeholder groups, or a combination of staff and stakeholders. Key candidates for the bureau are those community stakeholders who are willing to be the plan's ambassadors. The objective of the speakers bureau is to provide peer-to-peer and informal outreach for the plan. Speakers would receive key talking points, a presentation template, and training on desired presentation approaches as well as plan objectives.

Roving workshops are mini-workshops that are conducted in the community at places where people tend to congregate—for example, churches, schools, grocery stores, parks, and community events such as parades or street fairs. The idea is to go to the public rather than trying to make the public come to you. The activities are designed to be interactive and fun for the public and easy to transport and manage for the planning team. They are intended to go beyond *exhibitions* (discussed earlier), which provide no or limited public input, and instead engage the public in a constructive dialogue with planners and each other.

Public Participation in CAP Implementation

Many if not most of the actions identified in the CAP will require individuals, families, and businesses to change their behavior. It is one thing to provide additional bicycle lanes in a community; it is another to get people to drive their cars less and ride their bicycles more. Changed behavior cannot be accomplished through a few public service announcements or brochures available at the local library. Successfully changing behavior requires a sustained and multipronged effort to continue engaging the public in the issue of climate change.

CAPs should contain a section that identifies how the public will be included in plan implementation. Long-term participation in plan implementation is an area that can often be best addressed through partnerships with local nongovernmental organizations. In some cases, funding can be provided to support these activities. For example, a bicycling advocacy organization is well positioned to carry out ongoing education and periodic community events. A local utility provider is also able to support energy conservation and efficiency programs through grants, rebates, and loan programs. Partnerships also offer the opportunity to sustain implementation through collaborative funding. Nongovernmental organization partners may be able to leverage or secure funds that are not available to government agencies and, in turn, support CAP implementation.

In most communities, reducing GHG emissions locally requires a change in business-as-usual behavior. This change warrants a shift in our existing and future day-to-day use of energy, fuel, and resources. Collaboration, education, and engagement will be essential to achieving reductions through retrofits to our existing social and built environments and to sustain or increase reductions over time.

Chapter Resources

Center for Research on Environmental Problems, *The Psychology of Climate Change Communication: A Guide for Scientists, Journalists, Educators, Political Aides, and the Interested Public* (New York: Columbia University, 2009). http://www.cred.columbia.edu/guide/. Provides information on how to communicate with the general public about climate change. Also addresses small group participation and behavior change. Useful in communities where understanding of climate change may be low.

James L. Creighton, *The Public Participation Handbook: Making Better Decisions through Citizen Involvement* (San Francisco: Jossey-Bass, 2005). This is a practical guide to designing public participation programs for public policy issues. It contains applied techniques, worksheets, checklists, and numerous examples. It also has an extensive list of tools for outreach, education, and participation.

ICLEI, *ICLEI Resource Guide: Outreach and Communications* (January 2009). http://www.icleiusa.org/action-center/engaging-your-community/outreach-and-communications-guide. "The Outreach and Communications Guide is a tool to help local governments effectively communicate climate information to their constituencies. The Guide contains an array of steps and methodologies for communication and outreach efforts, as well as a compilation of best practices from around the United States."

International Association for Public Participation (IAP2). http://ww.iap2.org/index.cfm. IAP2 is an international membership-based association that promotes and improves the practice of public participation. It provides numerous tools for developing and implementing public participation as well as training and access to the latest research on public participation.

Kathleen Les, "Engage Your Community in Bold Initiatives on Climate Change," *Western City* (May 2008). http://www.westerncity.com/Western-City/May-2008/Engage-Your-Community-in-Bold-Initiatives-on-Climate-Change/. Explains how to conduct an effective community engagement effort for developing and implementing a climate action plan and includes numerous examples from communities.

Chapter 4

⚭

Greenhouse Gas Emissions Inventory

The technical definition of a community greenhouse gas emissions (GHG) inventory is an accounting of GHGs emitted by a community to (and in some cases, removed from) the atmosphere over a period of time, usually a calendar year. The inventory provides the baseline or existing condition from which to measure progress toward a GHG reduction target. This approach of quantifying a problem is not novel for most communities. Transportation studies that quantify the amount of traffic on roadways or housing studies that quantify the housing stock and assess its affordability are just two examples of gathering quantitative data in support of planning.

A GHG emissions inventory can be likened to an assessment you might do to begin a weight loss plan. First, you may weigh yourself and do an overall health assessment. What is your weight, cholesterol level, and percentage of body fat? Then you determine the sources of your calories: pancakes for breakfast, 700 calories; a burger, fries, and soft drink for lunch, 900 calories; and pizza, salad, and a glass of wine for dinner, 1,100 calories. You will use the information gathered in this inventory to change the intake of food, fat, and cholesterol for a healthier future. In the same way a community needs to understand its current emissions level, how community choices affect that level, and how it can change.

Principles of GHG Emissions Inventories

At the community scale, inventories focus on identifying sources and estimating quantities of GHG emissions. The primary sources are motor vehicles that directly consume fossil fuel and industrial, commercial, residential, and governmental buildings and operations that consume fossil fuels and electricity (thus indirectly consuming fossil fuels if

Box 4.1
Conversion of Greenhouse Gas Emissions to Familiar Equivalents

1,000 metric tons (MT) of CO_2 emissions is approximately equivalent to the following:

- CO_2 emissions from 112,100 gallons of gasoline consumed
- CO_2 emissions from 2,300 barrels of oil consumed
- CO_2 emissions from the electricity use of 120 homes for one year
- CO_2 emissions from burning 5.4 railcars' worth of coal
- Annual CO_2 emissions of 0.0002 coal-fired power plants
- CO_2 emissions from 41,600 propane cylinders used for home barbecues
- Greenhouse gas emissions avoided by recycling 350 tons of waste instead of sending it to the landfill
- Carbon sequestered by 25,600 tree seedlings grown for 10 years
- Carbon sequestered annually by 210 acres of pine or fir forests
- Carbon sequestered annually by 10 acres of forest preserved from deforestation

Source: U.S. EPA Greenhouse Gas Equivalencies Calculator. http://www.epa.gov/cleanenergy/energy
-resources/calculator.html.

no alternative form of electricity production is available). There are a number of other sources of GHG emissions at the community level but transportation and energy sources emit the majority of GHG emissions in most communities. GHGs are not measured directly; instead they are calculated based on measures such as how much electricity and natural gas is used in the community, how much people drive, what types of vehicles and fuels are present, and how much waste is generated. Each of these can be measured or estimated in a community and then converted to GHG emissions using standardized tools and databases. Because GHGs are calculated in this indirect fashion, they should be considered estimates, not exact measures. Box 4.1 provides an example set of estimates relating GHG emissions to community actions. By following appropriate protocols, these estimates can be sufficiently accurate for climate action planning.

A typical GHG emissions inventory will report the total annual emissions attributed to the community and a breakdown of their sources. Table 4.1 is an example from the City of San Carlos that displays the

Table 4.1 An example of conversion factors used by the City of San Carlos to calculate community-wide GHG emissions

Quantity	Value	Notes
Standard unit conversions		
1 pound (lb)	0.0004536 metric tons (tonnes)	Engineering standard
1 short ton (ton)	0.9072 metric tons (tonnes)	Engineering standard
1 metric ton (tonne)	1.1023 short tons (tons)	Engineering standard
	2,204.62 pounds (lb)	
1 kilowatt hour (kWh)	3,412 Btu (Btu)	Engineering standard
1 therm	100,000 Btu (Btu)	Engineering standard
City of San Carlos—General Greenhouse Gas Conversions for baseline 2005 calculations		
1 kilowatt hour (kWh)	0.492859 lb CO_2e	PG&E 2005 emissions factor certified by the California Climate Action Registry.
1 MMBtu	53.05 kilograms (kg) CO_2e	PG&E CO_2e emissions factor for delivered natural gas, certified by the California Climate Action Registry and CEC.
1 vehicle mile traveled (VMT)	1.077 lb CO_2e	Average estimate calculated by dividing total CO_2e derived from EMFAC and CACP by total VMT. Individual calculations may vary from this average coefficient based on model year and vehicle class.
1 short ton landfilled waste	0.277 metric tons CO_2e	Average estimate calculated by dividing total emissions from landfilled waste derived from EPA's WARM model and CACP by total tons landfilled. Individual calculations may vary from this average coefficient based on type of waste landfilled and waste management practices.

Note: This list of common conversions is used throughout the San Carlos Climate Action Plan. The City of San Carlos—General Greenhouse Gas Conversions are average estimates of the greenhouse gases (GHGs) produced by a unit of natural gas, electricity, and VMT within the city of San Carlos in calendar year 2005. The purpose of these conversion estimates is to provide an estimate for the reader to visualize the GHG equivalent of activities within the city.

Abbreviations: MMBtu, million British thermal units; PG&E, Pacific Gas & Electric; CEC, California Energy Commission; EMFAC, motor vehicle emissions model; CACP, Clean Air and Climate Protection; VMT, vehicle miles traveled.

Source: City of San Carlos, Climate Action Plan (2009).

Table 4.2.   Example of community-wide greenhouse gas emissions inventory summary

Sector	Metric tons CO_2e/year	Percent of total
Residential energy	49,178	18.4
Commercial/industrial energy	54,619	20.4
Transportation	150,663	56.4
Waste	12,777	4.8
Total community-wide emissions	267,237	100

conversion factors used to calculate GHG emissions. These are a combination of standard unit conversions and conversion factors derived from local data specific to San Carlos. These factors are used to quantify the GHG emissions produced by a community. Table 4.2 is an example of a standard summary of total emissions attributable to a community. The table shows emissions in metric tons of carbon dioxide equivalents (CO_2e); as discussed in appendix A, not all GHGs are equivalent to CO_2 in their global warming potential so they are usually converted to the same units. The table also shows the breakdown by the four most common sectors: residential energy, commercial/industrial energy, transportation, and waste. In some inventories these sectors are further subdivided (e.g., separating commercial and industrial) and some communities also break down total emissions by fuel type (e.g., gasoline, natural gas, coal, etc.).

Local governments most commonly use two types of inventories: a community-wide inventory and a local government operations inventory. It is considered best practice to conduct community-wide inventories that include inventorying local government operations as a distinct subset of the total emissions.[1] Local government operations typically comprise between 3% and 8% of community-wide emissions. Some communities do not break out local government operations. Moreover, some communities choose to inventory *only* local government operations emissions. Usually this is because they are preparing a climate action plan (CAP) that includes only reduction targets and emissions reduction strategies for local government operations; these are often called municipal CAPs.

Current best practice suggests that communities should prepare a baseline GHG emissions inventory prior to the development of a CAP. Communities that are members of ICLEI or signatories to the U.S. Mayors Climate Protection Agreement commit to preparation of the baseline inventory. A community that is faced with the choice between doing a CAP without a GHG inventory and not doing a CAP at all should proceed with doing a CAP without an inventory. After all, a community does not need to know its specific GHG emissions to know that taking actions to improve energy efficiency and conservation and reduce fuel consumption are prudent cost savings measures that also provide community quality of life benefits. The GHG emissions inventory is a detailed, technical exercise that will take time and expertise beyond the capacity of some communities, but there are good reasons to undertake this task.

The most effective CAPs contain GHG emissions reduction strategies tied specifically to sources of GHG emissions identified and quantified in an inventory. This should make logical sense; if most of a community's emissions are generated from transportation sources then most of the emissions reduction strategies should be aimed at transportation policies and programs. The most effective CAPs will also quantify the expected GHG reduction benefits and contribution of the strategies toward the overall reduction target. According to the U.S. Environmental Protection Agency, communities use GHG inventories to do the following:[2]

- Identify sources of emissions and their current magnitude within an area
- Identify and assess emissions trends
- Establish a foundation for projecting future emissions
- Provide a basis for future reduction targets
- Set a benchmark for tracking progress toward a reduction target
- Quantify the benefits of proposed emissions reduction strategies
- Provide a basis for developing a CAP

Inventories form the basis for decision making; they should thus be transparent and easily reproducible and should follow established protocol. This will also ensure consistency and comparability with future updates and other inventories. There are five accounting and reporting principles to ensure that "GHG data represent a faithful, true, and fair account" of a local government's GHG emissions.[3]

1. *Relevance*: The GHG inventory should appropriately reflect the GHG emissions of the community and should be organized to serve the decision-making needs of users.
2. *Completeness*: All GHG emission sources and emissions-causing activities within the chosen inventory boundary should be accounted for. Any specific exclusion should be justified and disclosed.
3. *Consistency*: Consistent methodologies should be used in the identification of boundaries, analysis of data, and quantification of emissions to enable meaningful trend analysis over time, demonstration of reductions, and comparisons of emissions. Any changes to the data, inventory boundary, methods, or any relevant factors in subsequent inventories should be disclosed.
4. *Transparency*: All relevant issues should be addressed and documented in a factual and coherent manner to provide a trail for future review and replication. All relevant data sources and assumptions should be disclosed, along with specific descriptions of methodologies and data sources used.
5. *Accuracy*: The quantification of GHG emissions should not be systematically over or under the actual emissions. Accuracy should be sufficient to enable users to make decisions with reasonable assurance as to the integrity of the reported information.

Despite the adherence to these principles it is important to keep in mind that while GHG emissions inventories based on current best practices are quite robust, they are still estimates with a degree of error. Preparers should not spend energy trying to account for the last 1% of emissions or estimating to significant digits not warranted by the many assumptions and estimates that make up the inventory calculations.

The Basic Inventory Process

The process of preparing a GHG emissions inventory entails a number of decisions and procedural steps that have been codified through a variety of GHG emissions inventory protocols and related software developed by national and international organizations. The basic steps for conducting a GHG emissions inventory in most protocols are as follows:

1. Data collection
2. Emissions calculations and reporting
3. Emissions forecasting
4. Emissions reduction target setting

This book does not provide a step-by-step explanation of how to conduct a GHG emissions inventory as this requires too much detail (e.g., the Local Government Operations Protocol manual is over 230 pages). Communities should refer to the chosen protocol manual for guidance. Instead, the book answers the following questions:

1. Who will prepare the inventory?
2. What is the appropriate methodology or protocol?
3. How should a baseline year be established?
4. What is the scope of the inventory?
5. What is a GHG emissions forecast?
6. How are emissions reduction targets selected?

Preparing the Inventory

Communities have several choices as to who will prepare the GHG emissions inventory: local government staff members, community volunteers, college faculty members and students, and consultants have been the most common choices. These vary by cost, experience and aptitude, and accountability. Preparing an inventory is detailed, time-consuming work that requires strong math and logic skills, solid organizational capabilities, and willingness to deal with uncertainties and assumptions. If the CAP will have a regulatory role, the inventory will need to be as accurate as possible, thoroughly documented, and legally defensible. A community will have to consider all these factors and make the best choice for its own circumstances.

Inventory Methodology or Protocol

Protocols establish *what* will be measured in an inventory and *how* it will be measured. There are a variety of GHG assessment protocols for businesses, governments, individuals, and other organizations, and some adventurous communities have created their own strategies by mixing and modifying a variety of existing protocols. Although this may work for some communities, most communities should use the widely adopted, standard protocols. The most common protocols for communities are the Local Government Operations Protocol and the Community-Scale Greenhouse Gas Emissions Accounting and Reporting Protocol.[4] These pertain to the two types of emissions inventories mentioned previously, local government operations inventories and community-wide inventories, respectively.

Local Government Operations Protocol

The Local Government Operations Protocol (LGO Protocol)[5] was developed in 2008 through the collaboration of the California Air Resources Board, ICLEI, the California Climate Action Registry,[6] and The Climate Registry.[7] The LGO Protocol is a tool for accounting and reporting GHG emissions across all of a local government's operations and is intended for use by local governments throughout the United States, with future application anticipated in Canada and Mexico. The LGO Protocol is based on the *Greenhouse Gas Protocol: A Corporate Accounting and Reporting Standard* developed by the World Resources Institute and the World Business Council for Sustainable Development (WRI/WBCSD). The Protocol is meant to be a "program neutral" guidance document available for use by any local government engaging in a GHG inventory exercise. It brings together GHG inventory guidance from a number of existing programs, namely the guidance provided by ICLEI to its Cities for Climate Protection campaign members over the last 15 years, the guidance provided by the California Climate Action Registry and The Climate Registry through their General Reporting Protocols, and the guidance from the State of California's mandatory GHG reporting regulation.

Community-Scale Greenhouse Gas Emissions Accounting and Reporting Protocol

The Community-Scale Greenhouse Gas Emissions Accounting and Reporting Protocol (Community-Scale Protocol) is set for release by

ICLEI in early 2012. The Community–Scale Protocol is a tool for accounting and reporting community-wide GHG emissions. This new protocol is tailored for use in U.S. communities and clarifies many issues with the previous protocol (the International Emissions Analysis Protocol). Moreover, the new Community–Scale Protocol is designed to complement the LGO Protocol.

Greenhouse Gas Inventory Software

There are two basic options for using the protocols to conduct the inventory and deal with the necessary calculations. The first is to use the detailed information in the protocols on data needs, assumptions, transformations, and calculations to manually construct an inventory spreadsheet in one of the commonly available spreadsheet programs such as MS Excel or Apple Numbers. Anyone who chooses this route should be very comfortable with the software and with math. The protocols provide sufficient detail and direction to do this successfully. This approach gives more flexibility with adjusting the data and assumptions for the local context and the accessibility of using commonly available spreadsheet software.

The second option is to use a custom GHG emissions inventory software package. ICLEI's Clean Air and Climate Protection (CACP) software has become the industry standard for GHG inventories. However, access to the software requires ICLEI membership. The CACP software was developed by ICLEI's State and Territorial Air Pollution Program Administrators and the Association of Local Air Pollution Control Officials (SAPPA/ALAPCO), and Torrie Smith Associates. It is based on the LGO Protocol and the International Emissions Analysis Protocol (soon to be the Community–Scale Protocol). The software enables communities to track and quantify emissions outputs and develop emissions scenarios to inform the planning process. The software calculates emissions resulting from energy consumption, vehicle miles traveled, and waste generation. The CACP software calculates emissions using specific factors (or coefficients) according to the type of fuel used. CACP aggregates and reports the three main GHG emissions (CO_2, CH_4, and N_2O) and converts them to equivalent carbon dioxide units, or CO_2e. The advantages of using the CACP software include ease of startup and access to ICLEI technical support and training.

There are a variety of third options for protocols and software, but these are not recommended. Some local governments, nonprofit organizations, and private sector businesses have prepared GHG inventories using their own methodology. For example, the nonprofit organization Clean Air–Cool Planet has a Small Town Carbon Calculator that can be used by small local governments to calculate local government operations emissions only.

Choosing Protocols and Software

In 2009, the authors reviewed a national sample of local CAPs and their associated GHG emissions inventories from a variety of U.S. cities. The available ICLEI protocols were used in 87% of the plans, thus making them the most popular choice. Of the plans that did not use the ICLEI protocols, some chose methods that closely followed the protocols and some chose to develop their own protocol that allowed the emissions inventory to more closely mirror the jurisdiction's view of global warming and their contribution to it. Aspen, Colorado, in particular, embodied the latter approach.[8]

Selection of the protocol, methodology, and software will depend on the purpose of the inventory and the resources available to the jurisdiction. The decision will depend on city staff support, budget allocations, time available, and availability of other resources or services such as consultants, volunteers, nonprofit organizations, or college faculty and students. Consultants or other organizations may develop tools and/or software tailored to the jurisdictions' needs. In addition, local governments must consider any regulatory mandates or guidance from state or regional agencies when applicable.

Establishing a Baseline Year

The GHG emissions inventory requires the choice of an inventory year, usually referred to as the baseline year. Communities should select the most recent calendar year for which consistent, comprehensive, and reliable data can be collected. The LGO Protocol recommends that local agencies select a baseline year that is "typical" and not a year in which emissions were influenced by unusual conditions such as extremely high or low economic growth, abnormal weather, or other outliers. Other issues may include whether the community wants to be consistent with state or neighboring jurisdictions' baseline years and consideration of the reduction target base year, which should be consistent. In addition, if a community has initiated several emissions reduction strategies, they may want to choose a year sufficiently in the past that recent GHG emissions reduction strategies could be counted toward reaching their emissions reduction target.

Inventory Scope

The typical sectors of a GHG inventory are residential and commercial energy, waste, and transportation. Other sectors, depending on the community, may include agriculture or industrial sources. The six greenhouse gases that should be quantified from these sectors and included in a GHG emissions inventory are carbon dioxide (CO_2), methane (CH_4), nitrous oxide (N_2O), hydrofluorocarbons (HFCs), perfluorocarbons (PFCs), and sulfur hexafluoride (SF_6). Other GHGs may be inventoried; however, methodologies for other GHGs may not be in the most commonly used protocols. According to the common protocols, emissions of CO_2, CH_4, and N_2O from fossil fuel combustion, electricity generation (the indirect emissions associated with electricity used in the community), waste disposal, and wastewater will be the most significant sources of GHG emissions in community-wide and local government operations inventories. Table 4.3 shows an example from the City of Richmond, California, of a detailed breakout of community-wide emissions and the data sources.

Inventories must be clear about the sources of emissions included and excluded, as these sources will form the basis of reduction measures in CAPs. The inventory is conducted by compiling activity data describing energy and fuel use and waste generation (see table 4.4 for typical activity data sources) and multiplying the activity data by emission factors for each type of energy used and each waste disposal site and technology. Protocol methodologies direct the application and selection of emission factors. Emissions are reported in terms of activity data, metric tons of each GHG, and metric tons of carbon dioxide equivalent (CO_2e). Converting emissions of non–CO_2 gases to units of CO_2e

Table 4.3 City of Richmond, California, 2005 greenhouse gas emissions inventory

Sector	Emissions source	Equiv CO_2 (metric tons)	Equiv CO_2 (%)	Energy (MMBtu)	Data source
Residential	Electricity	39,447	0.7	575,356	PG&E
	Natural gas	86,671	1.5	1,620,510	PG&E
Subtotal residential		*126,118*	*2.2*	*2,195,866*	
Commercial/industrial					
"District" direct access	Electricity	28	0.0	294	PG&E
Commercial/industrial PG&E	Electricity	79,392	1.4	1,157,964	PG&E
	Natural gas	1,517,105	25.9	28,365,648	PG&E
Other direct access	Electricity (estimated)	13,154	0.2	139,472	CEC
	Natural gas	8,907	0.2	166,536	From industry
BAAQMD monitored point source emissions		3,522,986	60.2		BAAQMD
Subtotal commercial		*5,141,572*	*87.8*	*29,829,914*	
Transportation					
Local roads AVMT	Gasoline	178,743	3.1	2,479,952	CalTrans
	Diesel	23,465	0.4	282,911	CalTrans
State highways AVMT	Gasoline	262,399	4.5	3,640,630	MTC
	Diesel	34,447	0.6	415,320	MTC
Rail	Diesel	7,788	0.1	92,723	BNSF/ Richmond
Subtotal transportation		*506,842*	*8.7*	*6,911,536*	
Waste					Pacific
ADC	Plant debris	804	0.0		CCC / CIWMB
Total waste disposed (w/o ADC)	Paper products	25,313	1.1		CCC / CIWMB
	Food waste	9,961	0.4		CCC / CIWMB
	Plant debris	2,664	0.1		CCC / CIWMB
	Wood/textiles	7,437	0.3		CCC / CIWMB
West Contra Costa Sanitary Landfill	Waste-in-place	32,309	0.6		CCC / Republic Services/ BAAQMD
Subtotal waste		*78,488*	*1.3*		
Grand total		**5,853,020**	**100.0**	**38,937,316**	

Abbreviations: ADC, alternative daily cover; AVMT, annual vehicle miles traveled; BAAQMD, Bay Area Air Quality Management District; BNSF, Burlington Northern Santa Fe; CalTrans, California Department of Transportation; CCC, Contra Costa County; CEC, California Energy Commission; CIWMB, California Integrated Waste Management Board; MMBtu, million British thermal units; MTC, Metropolitan Transportation Commission; PG&E, Pacific Gas & Electric.

Source: City of Richmond, *2005 Greenhouse Gas Emissions Inventory* (2009).

Table 4.4 Example of greenhouse gas emissions sectors, units of measurement, scope, and data source

Sector	Information	Unit of measurement	Emissions scope	Activity data source
Residential	Electricity consumption	kWh	Scope 2	Local utility provider
	Natural gas consumption	Therms	Scope 1	Local utility provider
Commercial and industrial	Electricity consumption	kWh	Scope 2	Local utility provider
	Natural gas consumption	Therms	Scope 1	Local utility provider
Transportation	Local road VMT	Annual average VMT	Scope 1	State database or local travel travel model
	Highway and interstate VMT for SLO County	Annual average VMT	Scope 1	State database or local travel model
Solid waste	Solid waste tonnage sent to landfill from activities in county	Short tons	Scope 3	Local landfill operator(s) or state reports
Off-road equipment	Emissions from off-road equipment and vehicles	Tons/year of N_2O, CO_2, and CH_4	Scope 3	State model or local estimates
Agriculture	Emissions from cattle and sheep	Head of cattle	Scope 3	County crop report
	Emissions from fertilizer use	Pounds of nitrogen	Scope 3	County crop report
Aircraft	Emissions in the landing and take-off operations (LTOs) zone	Grams of N_2O, CO_2, and CH_4	Scope 3	Local airport operator/ aircraft operations study

Abbreviations: CH_4, methane; CO_2, carbon dioxide; kWh, kilowatt-hours; N_2O, nitrous oxide; VMT, vehicle miles traveled.

provides comparison of GHGs on a common basis (i.e., on the ability of each GHG to trap heat in the atmosphere). Non-CO_2 gases are converted to CO_2e using internationally recognized global warming potential (see appendix A).

Emissions are quantified and tracked separately, and the results are presented in inventory reports by sector, source, and scope. Differentiating between emission scopes (Scopes) helps to avoid the possibility of double counting and misrepresenting emissions when reporting but allows all policy relevant information to be captured. In addition, tracking emissions sources separately allows decision makers to tailor reduction strategies. Three classifications, Scopes 1, 2, and 3, are used to categorize emissions sources, differing slightly when applied in the context of government operations and community-scale inventories (see table 4.3 for examples). Scopes are defined by the LGO Protocol as follows:

- Scope 1 emissions are all direct GHG emissions (with the exception of direct CO_2 emissions from biogenic sources). These are primarily emissions from motor vehicles and stationary sources such as power plants or factories located *within* the community.
- Scope 2 emissions are indirect GHG emissions associated with the consumption of purchased or acquired electricity, steam, heating, or cooling from power plants located *outside* of the community.
- Scope 3 emissions are all other indirect emissions not covered in Scope 2, such as emissions resulting from the extraction and production of purchased materials and fuels, transport-related activities in vehicles not owned or controlled by the reporting entity (e.g., employee commuting and business travel), outsourced activities, waste disposal, and so forth.

In standard protocols only Scope 1 and 2 emissions are used. This is partly a data issue. Data for Scope 3 emissions are difficult to obtain and their accuracy is questionable. Also, Scope 3 emissions are more economically and culturally complicated and less amenable to emissions reduction strategies (because they include things like household purchasing decisions and global manufacturing chains). There is increasing attention to overcoming these limitations for Scope 3 emissions, but the current practice is to focus on Scopes 1 and 2.

The designation of Scopes also has a relationship to the spatial boundaries of the inventory. Before collecting data, inventory preparers must identify spatial boundaries for the inventory and include all important sources of GHG emissions occurring within these boundaries. Community-wide inventories are typically based on the local government's political boundary. A jurisdiction may elect to inventory emissions outside of its political boundary. The most common reason is to be consistent with a comprehensive plan that may establish planning area boundaries for future land use beyond the community's current political boundaries. If this is done, it should be explained in the inventory.

The standard use of political boundaries for community-wide emissions can be confusing when one is considering Scope 2 emissions. In some communities electricity is produced within the political boundaries of the jurisdiction, which would be a Scope 1 emission source, but in most it is produced outside of the community and is thus a Scope 2 emission. Either way, the activities that create the demand for the electricity—powering homes and businesses—do occur within the political boundaries of the jurisdiction and should be inventoried. This distinction of Scope 1 and 2 emissions and political boundaries is certainly a debatable area of GHG emissions inventory practice, so whatever choice a community makes about this should be documented and justified.

Local government operations emissions include emissions arising from the use and operations of all facilities, buildings, equipment, and activities that are owned, operated, or managed by the local government. These are usually within the political boundaries of the jurisdiction, but some may exist outside the jurisdiction. For example, a local government may own a landfill or a water supply and transmission pipes outside its political boundaries. Since all emissions that are a consequence of the local government's operations must be included, these types of facilities and operations would be included. The LGO Protocol provides clear guidance on criteria for operational or financial control as stated in the LGO Protocol:[9]

> A local government has operational control over an operation if the local government has the full authority to introduce and implement its operating policies at the operation. One or more of the following conditions establishes operational control:
> - Wholly owning an operation, facility, or source; or
> - Having the full authority to introduce and implement operational and health, safety and environmental policies (including both GHG- and non-GHG- related policies).

A local government has financial control over an operation for GHG accounting purposes if the operation is fully consolidated in financial accounts. The LGO Protocol strongly recommends the use of the operational control approach to defining a jurisdiction's boundary for the local government operations inventory.

Greenhouse Gas Emissions Forecast[10]

It is common practice to prepare a GHG emissions forecast once the inventory has been completed. GHG forecasts are projections of possible future GHG emissions from all sectors of the inventory. Local forecasts for population, jobs, and housing are used to develop a forecast of future emissions; this is referred to as the business-as-usual (BAU) forecast (fig. 4.1). The BAU forecast can be thought of as what the emissions would be in the future if the community did nothing new to try to reduce them. The development of the BAU forecast is accompanied by the setting of a GHG emissions reduction target for the forecast year. The difference between the likely increase in emissions estimated in the BAU forecast and the emissions reduction target establishes the amount of emissions reduction that must be accomplished through strategies in the CAP. This is sometimes called the reduction wedge due to its appearance when graphed (see fig. 4.1).

Selection of Forecast Year(s)

During the selection of a baseline year, it is common practice for a municipality to select one or more forecast years. Forecast years are usually at least five years from the baseline year, with ten to twenty years being common. The first principle is that the forecast years should be consistent with the emissions reduction target years (described later in this chapter). Other considerations include the availability of a jurisdiction's forecasts for population, jobs, and housing, and the relationship of the inventory to other long-range planning documents or regulations. For example, most municipalities in California use 2020 as a forecast year to be consistent with goals of California's Global Warming Solutions Act and its subsequent implementation documents. The choice of a forecast year for the inventory also essentially establishes the planning horizon of the CAP.

Figure 4.1 Example of community-wide greenhouse gas (GHG) emissions forecast and reduction target. The upper line, business-as-usual (BAU), indicates the projected emissions if no additional actions to reduce emissions are taken. The second line, adjusted BAU, represents the local consequences of state and federal policy such as fuel efficiency regulations, renewable portfolio requirements, and energy efficient building standards. The bottom line represents a community's adopted GHG emissions reduction target. The wedge that is the difference between BAU and the reduction target represents the amount of emissions that must be reduced in a given community through a combination of local, state, and federal actions. The difference between the adjusted BAU and reduction target identifies the reductions to be achieved on a local level through implementation of a local climate action plan.

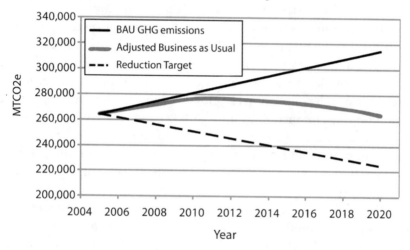

Adjusting the Forecast

Some GHG emissions forecasts show an adjusted BAU forecast. Since the forecasts for population, jobs, and housing used to develop the standard BAU forecast are typically simple extrapolation models based on historic data, they do not capture more complex factors that may affect these community measures. These factors are usually referred to as external (or exogenous) factors since they are not accounted for within the simple extrapolation model.

There are several types of external changes that will affect future levels of GHG emissions in a community: technological, social/behavioral, legislative and regulatory, demographic, and economic. Issues for technological innovation and change include automotive technology and fuels, electricity generation and fuels, and building technology. Social/behavioral changes may include commuting habits, household energy use, or purchasing habits. Potential legislative and regulatory change may include cap-and-trade legislation, renewable energy portfolio standards, and fuel efficiency standards (e.g., the Corporate Average Fuel Economy [CAFE] federal standard). Demographic changes that may influence GHG emissions include population growth, poverty level, and housing tenure and occupancy. Long-term GHG emissions may also be influenced by economic changes in gross domestic product, industrial and manufacturing mix, and balance of trade. This sampling of issues shows that considerable uncertainty exists in forecasting future levels of GHG emissions, particularly at the community level.

Two common solutions when dealing with uncertainty in forecasting are either to ignore it and use the original BAU forecast, or to develop multiple forecast ranges or scenarios. The problem with the former is that change seems almost certain at this point. For example, public transit ridership is at its highest level in fifty-two years,[11] bicycle commuting has jumped 43% since 2000,[12] and solar and wind power has consistently increased since 1999.[13] Emissions forecasts that assume long-term trends will persist and do not take into account the potential for dramatic changes over the short term may have important policy consequences.

The policy implications of ongoing external change that reduces GHG emissions could include the setting of overly conservative reduction targets, "sticker shock" reactions to how much effort would be required to meet aggressive reduction targets, or despondency created from a sense that the future growth of emissions is inevitable. Additionally, assuming no external changes puts communities in the position of misjudging the level of local GHG emission reductions needed. Too little reduction and the community misses its reduction target. Too much emissions reduction and the community may incur high costs (an economically inefficient outcome) or bear an unfair share of state and national reduction targets. Yet this is the most common assumption made. Most GHG emissions forecasts assume no external change due to the difficulty of forecasting this information.

Three recent studies show that external changes in technology and legislation have considerable effects on a community's choice about which emissions reduction policies to choose and how aggressive they need to be. The first study shows that in the Puget Sound region an "aggressive" set of assumptions about future changes in fuel efficiency (a 287% increase in fuel economy fleetwide) would still require local actions to reduce vehicle miles traveled (VMT) by 20%.[14] The second study estimates that half of San Diego's GHG emissions reduction target would have to be achieved through local measures and half through state requirements on utility renewable energy portfolio standards and low-carbon fuel standards.[15] The third study states that carbon neutrality for a university campus is a "fantasy unless there are supportive energy, transportation, and carbon sequestration initiatives at the state, national, and international level."[16]

The problem with addressing uncertainty by developing multiple forecast ranges or scenarios is that making assumptions about future changes in the areas listed would likely exceed the capability of most local governments. Moreover, no standardized approach for addressing this has been developed for community-level emissions inventories.

The issue of external change is one of the most difficult technical issues in GHG emissions forecasting. Guidance is poor and often conflicting; moreover, the rapid rate of technical, legislative/regulatory, and social change makes it challenging to adjust BAU forecasts to account for change. Yet these changes will have a significant impact on the ability of communities to develop mitigation actions to adequately account for their share of needed GHG emissions reductions. In fact, some communities are counting on this external change to help them achieve their targets. The best current advice is for communities to identify future emissions reductions resulting from statewide policies such as requirements for vehicle fuel mileage or targets for electricity production from low-carbon fuel sources. If the state has a CAP (see fig. 1.6) these types of policies may be specified, and they may be linked to forecasted GHG emission reductions. Since federal policy appears uncertain at this time, and the rate of technology change is very difficult to predict, it is not recommended to adjust the BAU based on these factors. In the future, better assistance from federal and state agencies on this issue is needed.

Selection of Emissions Reduction Targets

The emissions reduction target is the quantity of GHG emissions the jurisdiction wants to reduce by the forecast year. The reduction target is typically expressed as the percentage by which emissions will be reduced relative to a baseline year (e.g., 15% reduction from the 2005 baseline level by 2020). A jurisdiction may select more than one reduction target. Reduction targets may be short term, midterm, or long term or all three. The period will influence the range of actions and policy options used to achieve them. A local government may set a long-term goal but also have short-term targets that serve as incremental steps toward that goal. Target setting may include consideration of targets adopted by other levels of government, peer communities, feasibility of achievement or implementation, scientific studies and reports, and the urgency of the issue. Separate baseline years, target years, and reduction percentages may be established for local government operations and community-wide emissions.

Adopted Reduction Targets

Many communities choose to adopt reduction targets that have been established by other organizations or government agencies. The benefit is that these targets have usually been vetted scientifically, they relieve the community of having to develop their own analysis and standard setting, and they create consistency among communities. The downside is that they may not adequately capture local conditions and contexts and may not reflect local values.

International Standards

At this time the most notable international attempt to establish GHG emissions reduction targets has been the Kyoto Protocol. It is a protocol of the United Nations Framework Convention on Climate Change (UNFCCC) that was established in December 1997 in Kyoto, Japan. The Kyoto Protocol has been ratified by 187 countries but not the United States. The Kyoto Protocol establishes a total reduction of 5.2% below 1990 levels by 2012 for industrialized nations, but the exact standard differs by country. The United States was designated for a reduction of 7%. Despite the fact that the United States has not adopted the Kyoto Protocol, many U.S. communities have adopted the reduction target of 7% below 1990 levels by 2012. Of course, 2012 is upon us so the Kyoto Protocol reduction target is no longer useful. There have been a series of international meetings to update the Kyoto Protocol but they are yet to produce a new standard.

In 2007, the European Union agreed to a reduction target of 20% below 1990 levels by 2020. This agreement included additional targets of "20% of EU energy consumption to come from renewable resources" and "a 20% reduction in primary energy use compared with projected levels, to be achieved by improving energy efficiency." Collectively these targets are known as the 20-20-20 targets.[17] Although this is not an international standard, it does represent the policy of a large portion of developed nations.

Carbon-Based and Warming-Based Standards
Several organizations have proposed standards based on maximum levels of CO_2 in the atmosphere, which is linked to maximum average global warming, before serious consequences of climate change are evident. The problem with these standards is that they do not provide clear guidance or meaning at the local level. The 2010 Copenhagen Accord states the following:[18]

> To achieve the ultimate objective of the Convention to stabilize greenhouse gas concentration in the atmosphere at a level that would prevent dangerous anthropogenic interference with the climate system, we shall, recognizing the scientific view that the increase in global temperature should be below 2 degrees Celsius, on the basis of equity and in the context of sustainable development, enhance our long-term cooperative action to combat climate change.

In 2009, a group of climate scientists asserted that atmospheric CO_2 should not exceed 350 parts per million (ppm) ($\sim$ 450 CO_2e) if the 2°C threshold was not to be crossed.[19] As of July 2010 atmospheric CO_2 was 390 ppm.[20] Although it is difficult to translate these standards to a reduction target, research suggests that this would necessitate a 50 to 95% reduction from 1990 levels by 2050.[21,22]

National and State Standards
There are no official U.S. GHG emissions reduction targets, and states vary as to whether they have adopted standards (box 4.2).[23] As of 2010, a series of congressional bills has been proposed to establish U.S. targets within the context of a cap-and-trade program. The most recent and well known of these bills—neither of which has become law—are the

Box 4.2
National, Regional, and State Greenhouse Gas Reduction Targets

National-Level Targets

U.S. (provisional): 17% below 2005 by 2020[a]
Canada: 17% below 2005 by 2020[b]

Regional-Level Targets

Regional Greenhouse Gas Initiative (RGGI) (Members: Connecticut, Delaware, Maine, Maryland, Massachusetts, New Hampshire, New Jersey, New York, Rhode Island, and Vermont): Stable from 2009 to 2014 and 10% reduction from 2015 to 2018[c]

Western Climate Initiative (WCI) (Partners: Arizona, British Columbia, California, Manitoba, Montana, New Mexico, Ontario, Oregon, Quebec, Utah, Washington. Observers: Alaska, Colorado, Idaho, Kansas, New Brunswick, Nevada, Nova Scotia, Saskatchewan, Wyoming, Yukon, and in Mexico: Baja California, Chihuahua, Coahuila, Nuevo Leon, Tamaulipas, Sonora): 15% below 2005 by 2020[d]

U.S. State-Level Targets[e]

Arizona: 2000 levels by 2020, and 50% below 2000 by 2040

California: 2000 levels by 2010, 1990 levels by 2020, and 80% below 1990 by 2050

Colorado: 20% below 2005 by 2020 and 80% below 2005 by 2050

Connecticut: 10% below 1990 by 2020 and 80% below 2001 by 2050

Florida: 2000 levels by 2017, 1990 levels by 2025, and 80% below 1990 by 2050

Hawaii: 1990 levels by 2020.

Illinois: 1990 levels by 2020 and 60% below 1990 by 2050

Maine: 1990 levels by 2010, 10% below 1990 by 2020, and 75–80% below 2003 in the long term

Maryland: 25% below 2006 by 2020

Massachusetts: 80% below 1990 by 2050

Michigan: 20% below 2005 by 2025 and 80% below 2005 by 2050

Minnesota: 15% below 2005 by 2015, 30% below 2005 by 2025, and 80% below 2005 by 2050

Montana: 1990 levels by 2020

New Hampshire: 1990 levels by 2010, 10% below 1990 by 2020, and 75–80% below 2003 in the long term

New Jersey: 1990 levels by 2020 and 80% below 2006 by 2050

New Mexico: 2000 levels by 2012, 10% below 2000 by 2020, and 75% below 2000 by 2050

New York: 5% below 1990 by 2010, 10% below 1990 by 2020, and 80% below 1990 by 2050

Oregon: Stop the growth of GHG emissions by 2010, 10% below 1990 by 2020, and 75% below 1990 by 2050

Rhode Island: 1990 levels by 2010, 10% below 1990 by 2020, and 75–85% below 2001 in the long term

Utah: 2005 levels by 2020

Vermont: 1990 levels by 2010, 10% below 1990 by 2020, and 75–85% below 2001 in the long term

Virginia: 30% below business as usual by 2025

Washington: 1990 levels by 2020, 25% below 1990 by 2035, and 50% below 1990 by 2050

Note: Check with your nation, state, or region for updates.

[a] http://www.whitehouse.gov/the-press-office/president-attend-copenhagen-climate-talks
[b] http://www.climatechange.gc.ca/default.asp?lang=en&n=72f16a84-1
[c] http://www.rggi.org/home
[d] http://www.westernclimateinitiative.org/index.php
[e] http://www.pewclimate.org/what_s_being_done/in_the_states/emissionstargets_map.cfm

American Clean Energy and Security Act of 2009 (also known as the Waxman-Markey Bill) and the American Power Act of 2010 (also known as the Kerry-Lieberman Bill). Both would have established U.S. reduction targets of 17% reduction from 2005 levels by 2020 and 80% reduction by 2050. This 2020 target is also consistent with President Obama's pledge to the 2010 Copenhagen Accord, which is a follow-up to the Kyoto Protocol. Communities may want to consider adopting these targets since they are as close as we have currently come to a national standard.

Considerations for Adopting Local Reduction Targets

In a study completed in 2009 by the authors, the Kyoto Protocol target of 7% below 1990 levels by 2012 was set or exceeded by 50% of the studied communities. Of the communities that met the Kyoto Protocol target, not surprisingly, most cited the Kyoto Protocol as justification for their target. Of the communities that exceeded the target, most were not clear on why but some cited a desire to meet levels set by their peer communities. For example, the City of San Francisco cited their inspiration as the sixteen international cities that had formally declared their intention to exceed the Kyoto Protocol (through the Toronto Declaration communiqué to the Conference of Parties meeting in Morocco, November 2001).[24]

For the communities whose target was short of the Kyoto Protocol, some provided no clear justification for the target, whereas others cited feasibility of implementation of emissions-reduction strategies, an average of a variety of sources including similar communities, and state-adopted standards. In addition, Denver, Colorado, adopted a short-term per capita reduction target that allows for a significant increase in total GHG emissions over baseline.[25] The Denver plan (the only plan that allowed an increase) justified this increase by explaining that due to significant population growth this goal is "attainable." The Denver case highlights the challenge faced by fast-growing communities that will increase their emissions simply because they are adding people versus slow-growth communities that will see little increase over baseline even if they do nothing.

There are several additional examples that show the variety of choices and reasons for setting reduction targets. The City of Chattanooga, Tennessee, used a mix of external standards to set short-, medium-, and long-term emissions reduction targets. For 2012, they chose 7% below 1990 by 2012 to be consistent with the U.S. Conference of Mayors and Kyoto Protocol standard. For 2010, they chose 20% below 1990 by 2020 consistent with the European Union Council standard. And, for 2050 they chose 80% below 1990 as identified by the Intergovernmental Panel on Climate Change. The City of Cincinnati, Ohio, chose a very different route. Although they also established targets for different time periods (8% below 2006 by 2012, 40% below 2006 by 2028, and 84% below 2006 by 2050) they justified using analysis from their climate action planning process. Their justification for their targets included reductions to stabilize Earth's climate, goals established by other cities and counties, and practical, affordable reduction measures consistent with local objectives. Finally, the City of San Carlos, California, set goals (15% below 2005 by 2020 and 35% below 2005 by 2030) consistent with statewide GHG reduction targets established in the California Global Warming Solutions Act (Assembly Bill 32).

Including Science in a Climate Action Plan

The scientific and technical issues in doing GHG inventories raise the issue of whether and how to address the basic science of climate change within a CAP. Most CAPs have a chapter or section that explains the basics of climate change and GHG emissions. Although this is not required in a CAP, especially given that numerous references on the topic are available, it is common practice to include this information. The Chicago CAP has a graphically interesting two-page spread that provides a very brief primer on climate change. The CAP for San Francisco has twenty pages of climate change science, though mostly on the effects of climate change on the city.

A chapter or section on the science of climate change, if included, should answer the following questions for the reader:

1.   How do we know the planet is warming?
2.   What causes global warming?
3.   What are the consequences of global warming?
4.   What are the primary sources of anthropogenic greenhouse gases?

The answers to the last two questions should specifically address the local community to the degree possible. For example, it is not expected that global warming will affect all places equally on the globe with respect to heat waves, drought, rainfall, and the like. Nor do all communities produce the same amounts or types of GHGs.

The third question, on the consequences of warming, is difficult to answer for any specific locality given that the science to predict regional changes with confidence is only now emerging. A very useful starting point for understanding the potential local impacts of global warming is the 2009 report *Global Climate Change Impacts in the United States* produced by the U.S. Global Change Research Program.[26] The report documents the current and forecasted impacts of climate change in nine U.S. regions in the sectors of water resources, energy supply and use, transportation, agriculture, ecosystems, human health, and society. In addition to this report, many states have also prepared reports on the anticipated impacts of climate change. For example, the State of California produces a biennial *Climate Change Impact Assessment* report that identifies the impact of climate change on key sectors based on a variety of climate change scenarios. The fourth question, on the source of GHG emissions, can be answered locally through the preparation of a local GHG emissions inventory.

The purpose of the science included in a CAP will vary by community. In some communities the science simply serves to define the issues in the CAP. In other communities the science may be needed to inform a skeptical public of the need for climate action planning or to explain why the city has been motivated to act. Whatever the case, the CAP and associated education and outreach should explain to the community the purpose of knowing the science. Thinking this through will help the plan preparers focus on what content and what level of detail are actually needed.

Next Steps

Best practice standards for GHG emissions inventories are changing and improving on a regular basis. The choices and assumptions made in GHG emissions inventories, forecasts, and reduction targets influence selection and implementation of CAP policies and actions. Following the process outlined in this chapter is recommended to ensure effectiveness of the next steps in the climate action planning process.

Once the GHG emission forecast is complete and the reduction target is established, mitigation actions to reduce the community's GHG emissions are then developed and adopted. Adopted GHG reduction strategies must cumulatively reach the GHG emissions reduction target identified in the inventory.

Chapter Resources

General GHG Inventory Information for Local Governments

Environmental Protection Agency (EPA) State and Local Climate and Energy Program: Developing a Greenhouse Gas Inventory. The EPA provides general GHG inventory resources and examples to local governments at the following websites:

- http://www.epa.gov/statelocalclimate/local/activities/ghg -inventory.html
- http://www.epa.gov/statelocalclimate/local/local-examples/ghg -inventory.html

GHG Inventory Accounting Protocols

International Emissions Accounting Protocol (IEA Protocol). The International Emissions Accounting Protocol is available for free download from the ICLEI USA website.

- http://www.icleiusa.org/programs/climate/ghg-protocol
  /international-emissions-analysis-protocol

Local Government Operations Protocol for the Quantification and Reporting of Greenhouse Gas Emissions Inventories, Version 1.1 (May 2010). The California Climate Action Registry (CCAR), the California Air Resources Board (CARB), ICLEI–Local Governments for Sustainability (ICLEI), and The Climate Registry (The Registry) collaborated and developed the LGO Protocol in 2008. The LGO Protocol is available at no charge at the following websites:

- http://www.arb.ca.gov/cc/protocols/localgov/localgov.htm
- http://www.icleiusa.org/programs/climate/ghg-protocol/local
  -government-operations-protocol
- http://www.climateregistry.org/tools/protocols/industry-specific
  -protocols/local-government-operations.html
- http://www.theclimateregistry.org/resources/protocols/local
  -government-operations-protocol/

GHG Inventory Tools and Software

Campus Carbon Calculator. The Campus Carbon Calculator was created by Clean Air–Cool Planet and is intended for use by colleges and universities.

- http://www.cleanair-coolplanet.org/toolkit/inv-calculator.php. Clean Air Climate Protection (CACP) Software. ICLEI–Local Governments for Sustainability provides the Clean Air Climate Protection Software to members only. The software is based on the Local Government Operations Protocol and the International Emissions Accounting Protocol.
- http://www.icleiusa.org/action-center/tools/cacp-software

The Small Town Carbon Calculator (STOCC). The STOCC tool was created through the collaboration of Clean Air–Cool Planet, Carbon Solutions New England, and the University of New Hampshire. STOCC is intended for small towns with relatively few municipal buildings/facilities and vehicles.

- http://www.cleanair-coolplanet.org/for_communities/stocc.php

Chapter 5

————————— ✻ —————————

Emissions Reduction Strategies

Strategies to reduce greenhouse gas (GHG) emissions form the core of a climate action plan (CAP). The emissions reduction strategies (also frequently called mitigation strategies) are the actions, programs, and policies that a community undertakes to reach its GHG emissions reduction targets. Common examples include constructing new bicycle paths, providing incentives for installation of solar panels, and requiring that new development meet strict energy efficiency or "green" building standards. The development of these strategies is an iterative process that should balance the GHG reduction potential, upfront and ongoing costs, and social and political feasibility. Most reduction strategies have benefits beyond emissions reduction; these are called co-benefits. For example, reducing GHG emissions can also lower ground-level ozone concentrations in a community, which will yield public health benefits, especially for those who suffer from asthma or other respiratory conditions. The development of reduction strategies should be seen as an opportunity not only to reduce GHG emissions and the progression of climate change but as a chance to position a community to become more economically, environmentally, and socially sustainable.

Because GHG emissions result from a range of urban processes, operations, and behaviors, successful reduction of GHG emissions relies not only on governmental action but also on the commitment of community members and collaboration with business, industry, and community organizations. Many members of a community, including business and industry, have embraced green or sustainability principles and seek to be involved in efforts that show their commitment to a better environment. They bring resources, audiences, and ideas that local governments may not.

GHG emissions reduction is an increasingly common area of policy development. As a result, there are many resources that provide examples of successful reduction strategies. The challenge for local jurisdictions is identifying those strategies that best meet local needs. This chapter provides guidance for developing such reduction strategies. It does not provide a comprehensive list of emissions reduction best practices; these continually evolve and are available in other resources (see the resources listed at the end of this chapter). Instead, it identifies the key issues and decisions that must be addressed during reduction strategy development. Reduction strategy development builds on data collected through the emissions inventory, policy audit, and public participation process (fig. 5.1).

Developing the Emissions Reduction Strategies

One of the first tasks in developing reduction strategies is to organize the community partners and the public to establish a participation process as discussed in chapters 2 and 3 (part of Phase I in fig. 5.1). Regardless of how this is done, those teams that will brainstorm, develop, and finalize the reduction strategies must work through the following considerations (discussed in more detail in the text):

1. What are the key sectors of the community to target for the most effective and efficient reduction of GHG emissions?
2. What types and mixes of strategies will be the most appropriate?
3. What level of analysis will be conducted to estimate the GHG emissions reductions of the proposed strategies, if any?
4. How will the reduction strategies be evaluated?
5. What should a reduction strategy include?

Targeting Key Areas

One of the first steps in reduction strategy development is careful evaluation of the GHG emissions inventory (see chap. 4), policy audit (see chap. 3), and community characteristics. These provide data necessary to identify areas of focus for development of reduction strategies that will best meet the needs and capabilities of the community.

The sectors shown in the GHG emissions inventory that contribute most to local emissions must be targeted in the reduction strategies. For example, in a community where a large percentage of GHG

Figure 5.1 The iterative steps of greenhouse gas reduction strategy development within the overall climate action planning process. The various phases and steps reflect those introduced in chapter 2.

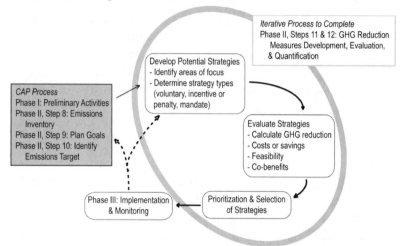

emissions come from a coal plant that produces electricity used in local buildings, a strategy should be to reduce electricity use and perhaps explore renewable energy sources.

The local policy audit, introduced in chapter 2, should be conducted around the time of the emissions inventory. The audit identifies community policies already in place that may support or be in conflict with reduction goals. For example, many communities already have programs to improve energy or water efficiency that also reduce local GHG emissions. Measures such as standards for historic building restoration or planned low-density development have the potential to conflict with emissions reduction goals. Current and pending national, state, and regional policy should also be included in the policy audit to evaluate changes outside of local control that may influence local emissions. These policies can include federal fuel efficiency standards for passenger automobiles or requirements for the percent of renewable energy supplied by energy providers (i.e., renewable energy portfolio standards). In 2009, the U.S. Environmental Protection Agency (EPA) announced that carbon dioxide is a threat to public health and welfare.[1] This finding may result in stronger and more direct action being taken at federal, state, and regional levels. This EPA finding may influence vehicular and industrial emissions regulation.

Finally, a set of basic community characteristics is important to complement these resources, including data such as the distribution of housing stock (age of structures and structure type); typical commute length; demographic and economic data such as age, income distribution, and housing affordability; and environmental data such as topography, temperature profile, wind patterns, and solar exposure. These data provide a strong basis for local strategy development and prioritization. The assumptions used for the business-as-usual forecast in the emissions inventory such as population, housing, and transportation growth rates serve a similar role in the formulation of reduction strategies. For example, slow-growth communities (e.g., < 1% per year) or communities where the building stock consists primarily of older structures will need to focus more specifically on retrofitting existing buildings to achieve energy efficiency improvements than a rapidly growing region that can more easily achieve efficiency improvements from standards on new construction. Community data can provide additional understanding of the inventory data. For example, a community that has a large incoming daily commute due to housing affordability can address two sources of emissions by focusing on local affordable housing. This reduces transportation emissions by reducing commute length and reduces residential energy use through construction of energy efficient housing.

In summary, a community should answer questions such as the following when beginning development of reduction strategies. These questions can be adjusted and/or supplemented depending on local conditions.

- What are the largest sources of emissions in the community? (GHG Emissions Inventory)
- What is the community already doing that reduces GHG emissions? (Policy Audit)
- What are the expected population, housing, and transportation growth/ decline rates of the community? (GHG Emissions Inventory)
- Are there community characteristics that influence emissions-generating behaviors? (Community Characteristics)
- What is the age of the housing stock? (Community Characteristics)
- What is the typical commute length (and is it more in-commuters or out-commuters)? (GHG Emissions Inventory and Community Characteristics)

- What is the topography, temperature profile, wind pattern, and solar exposure of the community? (Community Characteristics)

Types of Reduction Strategies

Strategies that serve to reduce GHG emissions take three forms: mandates, incentives or disincentives, and voluntary actions. The choice of strategy type must be made with careful consideration of local context. The policy audit can be helpful here since much can be learned from existing community strategies that have proven successful. This helps clarify the full range of policy options, which is important for the process of balancing necessary reduction areas against other community needs. Mandates may have higher costs or face greater political resistance, but more confidence can often be placed in the emissions reductions being realized. For example, a strategy that requires an energy efficiency building retrofit at the point of sale or major renovation is likely to be far more effective in reducing emissions than an incentive program that offers a small rebate to community members willing to voluntarily update household appliances to improve energy efficiency. But requiring such an upgrade will require more work, including assumptions to be made about the number of units sold or remodeled per year, and may meet some resistance. Mandates should also be evaluated with consideration being given to which members of a community are most likely impacted (e.g., bearing the bulk of the cost or excluded due to increased costs). Strategy types should also be combined, particularly for strategies that are politically challenging. Successful phasing of new mandates in the face of political opposition can begin with outreach and education to build community support for the strategy, followed by an incentive program to encourage voluntary compliance, and only then require the proposed changes. Many emissions-reducing strategies rely, in part, on voluntary behavior change. As a result, combining incentives and education with strategies such as the provision of new infrastructure bolsters long-term effectiveness.

Quantifying Greenhouse Gas Emissions Reductions

The quantification of anticipated GHG emissions reduction with each strategy facilitates the assessment of whether identified reduction targets are being met. Quantification is therefore one of the most important components of climate action planning and is regarded as the key criterion for evaluating reduction strategies. Unfortunately there are no standardized GHG emissions reduction estimates that can be assigned to reduction strategies. GHG reductions are estimated based on a variety of measures and assumptions that differ by community or region (boxes 5.1 and 5.2). For example, electricity in the Midwest is largely produced from coal, whereas in the West it is from natural gas and hydropower, resulting in very different GHG emissions reductions of energy efficiency strategies.

The most important step in quantifying reduction strategies is that the measures and assumptions are explicit and documented. This allows for monitoring, evaluation of implementation success, and identification of whether or not the initial assumptions continue to be accurate. An example of this transparency can be found in the City of Worcester, Massachusetts, CAP, which includes all constants and assumptions used to calculate emissions reduction in the appendix.[2] The emissions inventory is the source for key constants such as average fuel efficiency of vehicles or the average GHG emissions per kWh of electricity used. If implementation fails to produce the desired result, the City will be able to examine which assumptions were faulty.

Emissions reduction calculations can be detailed quantifications or rough estimates. A CAP will likely include both depending on the measure type, level of importance, and certainty with which assumptions can be made. In some cases, there will be high levels of uncertainty in estimating strategy effectiveness. This is particularly true for strategies that rely on voluntary community action. Due to high levels of uncertainty, conservative estimates or an estimate range may be enough. Also, strategies that comprise a very small portion of total GHG emissions reduction may not require precise estimates of GHG reductions. For example, an incentive program encouraging drought-tolerant landscaping will reduce GHG emission associated with the treatment and delivery of water. However, water often represents a very small portion of community emissions, and it is difficult to estimate the level of community participation in this program and the amount of water that will be saved. In this case, a simple set of assumptions that are locally appropriate such as percent participation and percent water use reduction per participating household is likely to be enough. If this strategy was a

Box 5.1
Examples of Greenhouse Gas Emissions Reduction Estimation

The City of Albany, California, adopted a reduction strategy to convert all street lights to LED bulbs or LED-solar systems. The following description from the plan explains the data and assumptions necessary to calculate the annual emissions reductions that would result from implementation of the strategy:[a]

> This measure is based on the energy efficiency of LED bulbs or LED solar systems with respect to the existing street light system. The GHG emission reduction potential of this measure was calculated conservatively assuming that all street lights would be converted to LED bulbs and not LED-solar systems. The energy savings associated with this measure were calculated assuming LED bulbs are 70% more energy efficient than the existing street lights. The City was able to provide total kilowatt-hours used for the existing streetlight system, to which the 70% reduction was applied. The GHG emission reduction associated with this measure was calculated using the same PG&E-specific electricity consumption emission factor used to calculate the City's GHG emissions associated with electricity consumption. In reality, this measure may have a greater GHG emission reduction potential due to the installation of solar systems in addition to the LED bulbs. Measure performance = 170 MT/year.

The description shows that the source metric is based on the assumption about the relative efficiency of LED bulbs (a universally valid assumption) and the electricity consumption data from the City. The emissions factor from the utility provider is based on the amount of CO_2e produced per kWh based on the utilities' fuel type. This is then combined with the global warming potential constant to calculate the GHG emissions reduction.

The City of Worcester, Massachusetts, adopted a policy to offer City employees the opportunity to telecommute one day per week.[b] The City assumed that one-eighth of employees would participate. They made a "drive-alone" assumption based on U.S. Census mode share statistics for the City and applied this to the number of City employees to estimate a potential number of drive-alone employees who would telecommute. This, combined with average commute length statistics from the GHG emissions inventory, allowed the City to calculate the source metric, which was the vehicle miles traveled (VMT) reduction resulting from the program. Combining this with the emissions factor from the emissions inventory (GHG/mi.), the City calculated the GHG emissions reduction.

[a] City of Albany, California, Climate Action Plan (April 2010), B-5.
[b] City of Worcester, Massachusetts, Climate Action Plan (December 2006).

Box 5.2

Comparing the Difference in Assumptions in Greenhouse Gas Emissions Reduction Estimation

Both the cities of Denver, Colorado, and Cincinnati, Ohio, included a car-share program in their climate action plans. In both cases the programs were seen as short-term strategies with emissions reductions calculated by 2012. The relative importance placed on the strategy within the plan and the data and assumptions based on local characteristics affected the estimation of the emissions reductions. In Cincinnati the estimated reduction was 5,000 tons CO_2e, and in Denver it was part of a suite of actions that resulted in reduction of 27,000 tons CO_2e.

Cincinnati, Ohio[a]

- The program was assumed to be implemented by an operator independent from, but supported by, the City.
- Locations focused on areas near the University of Cincinnati and the central business district (CBD). Anticipated membership from these locations was 1,000–3,000 for the university and 200–500 in the CBD.
- The anticipated membership was assumed to reduce regular use of 2,000–5,000 private vehicles.
- Taking the low end of this estimate (to be more conservative), 2,000 vehicles were assumed to reduce their mileage by half. Average annual miles driven was assumed to be 10,000.
- The estimated emission rate of 1 lb of CO_2e per mile yielded an annual reduction of approximately 5,000 tons of CO_2e.
- Initial investment to start the program was estimated to be $1,000,000, with a program maturity period of two to four years.

Denver, Colorado[b]

- The car-share program was included as one of a suite of programs labeled Alternative Transportation Strategies that included hybrid taxis, bicycling, walking, van/car pools, mass transit, and business support for bicycle commuting.
- The emissions reductions of all strategies were lumped together, but a suite of assumptions were made about the car-share program specifically.
- By 2012 it was assumed that 70 car-share vehicles would be in use.
- The program was assumed to be equivalent to taking 500 cars off the road.
- All alternative transportation strategies were estimated to yield a total reduction in emissions of 27,000 CO_2e, with the car-share program contributing a portion.
- Initial investment was identified as purchase of the 70 vehicles.

[a] City of Cincinnati, *Climate Protection Action Plan: The Green Cincinnati Plan* (2008), http://www.cincinnati-oh.gov/cmgr/downloads/cmgr_pdf18280.pdf.
[b] City of Denver, Climate Action Plan (2007), www.greenprintdenver.org/docs/DenverClimateActionPlan.pdf.

major portion of overall GHG reductions, it may be appropriate for factors such as evapotranspiration rates, wind, and average yard size to be considered in the reduction estimate.

One of the most complete references for quantifying reduction strategies is the August 2010 California Air Pollution Control Officers Association (CAPCOA) guide for local governments, *Quantifying Greenhouse Gas Emissions Measures*. The goal of the report is "to provide accurate and reliable quantification methods that can be used throughout California and adapted for use outside of the state as well."[3] The report contains a series of fact sheets on particular types of reduction strategies and accompanying guides on how to use the fact sheets. The quantification methods are based on using readily available data gathered by the planning team. The CAPCOA report describes the basic logic of emissions quantification:[4]

> The general equation for emissions quantification is shown here for each GHG:
>
> GHG Emissions = [source metric] $\times$ [emission factor] $\times$ [GWP]

where *source metric* and *emission factor* are defined as follows:

> Source metric: The "source metric" is the unit of measure of the source of the emissions. For example, for transportation sources, the metric is vehicle miles traveled; for building energy use, it is "energy intensity," that is, the energy demand per square foot of building space. Mitigation measures [reduction strategies] that involve source reduction are measures that reduce the source metric. This can include, for example, reducing the miles traveled by a vehicle because the reduction in miles traveled will reduce the emissions generated from vehicle travel. Similarly, a reduction in dwelling unit electricity use by installing energy efficient appliances and lighting will reduce the emissions associated with total electricity assigned to dwelling units.
>
> Emission factor: The "emission factor" is the rate at which emissions are generated per unit of source metric (see above). Reductions in the emission factor happen when fewer emissions are generated per unit of source metric, for example, a decrease in the amount of emissions that are released per kilowatt-hour, per gallon of water, etc. Such a decrease may apply if a carbon-neutral

electricity source (e.g., from photovoltaics) is used in place of grid electricity, which has higher associated emissions; or if electricity is used instead of combustion fuel, such as with electric cars. Reductions can also occur if a fuel with lower GHG emissions is used in the place of one with higher GHG emissions. From a quantification standpoint, for this type of measure, it is the "emission factor" in the equation that changes.

Evaluating Reduction Strategies

A community may identify many reduction strategies, but only some of them will be appropriate for inclusion in the CAP. Therefore, there should be a process to evaluate and prioritize each reduction strategy to ensure that it meets a community's needs and constraints (see box 5.2). Identification of local needs and constraints and establishment of the process necessary for this assessment begins with the formation of the Climate Action Team (chap. 2) and public outreach (chap. 3). Many CAPs now disclose the analysis of elements that contribute to prioritization of strategies. This information allows for greater transparency in the plan formulation process and more clearly sets a path for implementation. The questions listed here should be addressed through the work of the Climate Action Team and through a public participation process for each reduction strategy considered. Others can be added based on local need. These questions are interrelated. For example, the need for funding may delay strategy implementation, which will subsequently adjust the emissions reduction estimates.

- What is the potential emissions reduction that will result from strategy implementation?
- How long will it take to begin implementation of a strategy?
- How long will it take for a strategy to be fully implemented?
- How much does the strategy cost to implement (initial and ongoing costs)?
- What is the political and social feasibility of the strategy?
- Are there co-benefits to implementation?

What is the potential emissions reduction that will result from strategy implementation?
The amount of GHG emissions reduction possible from each strategy should be compared to assess the relative value of each. Calculations

should be made at the lowest level of strategy development without overlap or double-counting. There are strategies that cannot be separated from each other. For example, bicycle infrastructure including paths, lighting, and storage may all serve to increase the community ridership; however, this increase cannot be divided between implementation of these items. The increase in ridership occurs from the collective impact of all of these items.

How Long Will It Take to Begin Implementation of a Strategy?
In some cases, a measure can be universally hailed by staff, the community, and advisory bodies but still require a series of actions to be completed prior to the start of implementation. This can be as simple as the time it takes to update or amend the comprehensive plan or drafting a new ordinance. In other cases, this may involve securing funding through grants or a local fee system to initiate a program. Each measure should be evaluated for how soon it can realistically be implemented.

How Long Will It Take for a Strategy to Be Fully Implemented?
In the context of emissions reduction estimates, full implementation refers to the time it will take to achieve the estimated emissions reduction. For example, a new ordinance can be adopted in the short term. This implements the measure; however, experience of the subsequent emissions reductions will be distributed through time. In some cases, a measure will identify a change that will take many years to fully achieve. For example, a green building ordinance aimed at new residences in a slow-growing community will likely produce benefits at a much slower rate than a strategy aimed at retrofitting existing residences.

How Much Does the Strategy Cost to Implement
(Initial and Ongoing Costs)?
The high upfront cost is one of the biggest limitations of climate-friendly strategies such as building retrofits for energy efficiency, renewable energy, or vehicle upgrades. This is a critical factor for evaluating and planning for measure implementation. If a funding mechanism is not identified for a strategy, time to raise necessary funds must be planned into the phasing of strategies. The initial costs of a measure can be a critical consideration when prioritizing strategies for immediate implementation. Given the limited budgets of local governments, funds must be carefully allocated.

In addition to initial costs, many projects carry ongoing costs of implementation, including materials, maintenance, and administration. Accurate estimation of these costs and a way to raise these funds are both vital to reduction measure formulation. Often these costs can be covered through adjustment in fee structures such as more aggressive tiered pricing for water or establishment of a fund to hold impact fees from new development. However, adjustments in fees must also be accompanied by an evaluation of which populations or community members will be most impacted. This addresses the issue of who bears the costs and who receives the benefits. This can be difficult politically and raises issues of fairness and social justice. Communities should ensure that the costs of strategy implementation are not unfairly borne by a narrow sector of the community, especially those least well-off. In West Hollywood, California, for each emissions reduction strategy the CAP contains magnitude estimates of cost to the City and private costs and savings that accrue to individuals or businesses.[5]

Cost-effectiveness refers to a comparison of the costs and benefits (and possibly co-benefits). In climate action planning it is common to see emissions-reduction strategies evaluated on their dollar cost per ton of GHG emissions reduced. For example, Denver estimated that their Residential Climate Challenge would cost $10–$26 per metric ton of CO_2e emissions reduced. Comparing strategies on cost-effectiveness can allow communities to identify how to get the biggest bang for the buck.

What Is the Political and Social Feasibility of the Strategy?

Awareness of potential political challenges is important during measure formulation. These difficulties can be addressed in a variety of ways, such as direct engagement with concerned stakeholders to devise a more palatable approach, including outreach and education as part of measure implementation, and careful choice of wording to avoid pitfalls. Strategies to identify community priorities and concerns are discussed in chapter 3.

There should also be consideration of how responsive the "target" will be to the strategy. For example, an ad campaign to get people to drive less may be fairly simple to create, but it may be a difficult way to successfully realize the emissions reduction since it requires people to change ingrained behaviors. On the other hand, having a city council raise parking meter rates will assuredly result in a change in parking usage and revenues, although this may meet political opposition. Some communities refer to this issue as "ease of implementation."

Communities are unique. Some climate strategies will fit right in with the current ethos of a community; others might be seen as radical. For example, new bicycle initiatives or public spending on bicycle infrastructure is likely to be easily welcomed in Boulder, Colorado, a city that takes pride in its bike culture. A strategy aimed at creating green jobs is likely to be similarly welcomed in communities with high unemployment rates.

Are There Co-benefits to Implementation?

Climate action planning is just one aspect of community planning and can be viewed as an opportunity to meet a variety of local goals. Strategies that carry benefits beyond mitigating climate change are most easily promoted to the public as well as decision makers. Categories of co-benefits include the following:

- Cost savings
- Energy conservation
- Health benefits
- Local business support
- Municipal revenue enhancement
- Water conservation
- Education and awareness
- Mobility improvement
- Climate adaptation
- Smart growth
- Water and air quality improvement
- Green space and recreation improvement
- Quality of life improvement
- Job creation
- Community development and redevelopment

Of these, cost savings and energy conservation are the most common desired co-benefits. In fact some communities organize their CAPs around these as their primary benefits and treat GHG emissions reduction as the co-benefits.

Prioritizing the Strategies

Once the evaluation is complete there are several methods for using the results to rank the strategies. The most straightforward is to pick a single criterion, then choose the strategies that perform best on that criterion. For example, some communities prioritize cost-effectiveness; they are looking for the greatest emissions reduction for the least cost, whereas others have identified co-benefits as the most important strategy. The challenge arises when multiple criteria are used to rank strategies. In this case each strategy can be scored on each criterion, then the scores can be added up and strategies ranked based on the best overall scores. The scores can be weighted by adding bonus points or multipliers to the most important criteria.

Contents of a Reduction Strategy

An emissions reduction strategy should contain enough detail that it can be implemented; it should be written as more than a goal. For example, a CAP may state: "increase transit ridership by 5%." This is a good goal, but it is not a strategy because it does not contain enough information or describe how the goal will be met. Instead, the CAP should contain a specific set of actions that if taken would result in increasing ridership by 5%, such as a marketing campaign, fare reductions, or routing changes.

In addition to specifying actions, each reduction strategy must include five pieces of information: an estimate of GHG reductions, a funding source, a phasing plan (how soon can it be implemented and how long will that take), an entity or department responsible for implementation, and identification of an indicator that will allow for effectiveness to be monitored. These are discussed in more detail in chapter 7.

Emissions-Reduction Strategy Sectors

Reduction strategies are usually organized in sectors similar to the emissions inventory. This chapter uses the sector approach with the addition of several that are generally not addressed in the inventory because they do not emit GHGs, such as sequestration or renewable energy. The sectors covered are transportation and land use, energy efficiency, renewable energy, carbon sequestration, agriculture, industry, waste, green living, and offsets. Embedded in each of these sectors are strategies that can be specifically targeted at local government operations. If a community sets a separate target for local government operations, it may be appropriate to move all strategies that specifically address government strategies to a separate chapter.

Local Government Operations

Communities often choose to lead by example. This can take the form of adopting more aggressive targets for local government operations or more aggressive implementation plans for reduction strategies that specifically target local government operations and employees. Many communities choose to do this because these are the aspects of the community over which local policy has the most control. A community government can simply move to more fuel efficient fleet vehicles as part of standard turnover much more easily than it can devise policy that would result in a community-wide move to improved fuel efficiency. Despite the differences in ease of implementation the emissions sectors in which strategies can be devised are consistent with those for community-wide reduction strategies. For this reason, the discussion of sectors does not distinctly break out local government operations.

Transportation and Land Use

In the United States, transportation accounts for about 27% of all GHG emissions. These emissions are from the combustion of fossil fuel, including gasoline in personal vehicles, diesel fuel in heavy-duty vehicles, and jet fuel in aircraft. The *Moving Cooler* report, prepared by the Urban Land Institute, identifies four basic approaches to the reduction of GHG emissions from the transportation sector: vehicle technology, fuel technology, vehicle and system operations, and travel activity.[6] The first of these two areas largely fall outside of the influence of communities since they are tied to state and federal policy and funding and the evolution of technology, so the focus is on the latter two. The necessary reductions in transportation emissions will require a variety of approaches. *Moving Cooler* states the following:

> The United States cannot reduce carbon dioxide (CO_2) emissions
> by 60 to 80 percent below 1990 levels—a commonly accepted

target for climate stabilization—unless the transportation sector contributes, and the transportation sector cannot do its fair share through vehicle and fuel technology alone. . . . The increase in vehicular travel across the nation's sprawling urban areas needs to be dramatically reduced, reversing trends that go back decades.[7]

To address the issue of vehicular travel—usually measured as the total or average number of vehicle miles traveled (VMT) in a community—the *Moving Cooler* report recommends development of reduction strategies in nine areas:[8]

- *Pricing and taxes*: Raise the costs associated with the use of the transportation system, including the cost of vehicle miles of travel and fuel consumption.

- *Land use and smart growth*: Create more transportation-efficient land use patterns, and by doing so reduce the number and length of motor vehicle trips.

- *Nonmotorized transport*: Encourage greater levels of walking and bicycling as alternatives to driving.

- *Public transportation improvements*: Expand public transportation by subsidizing fares, increasing service on existing routes, or building new infrastructure.

- *Ride-sharing, car-sharing, and other commuting strategies*: Expand services and provide incentives to travelers to choose transportation options other than driving alone.

- *Regulatory strategies*: Implement regulations that moderate vehicle travel or reduce speeds to achieve higher fuel efficiency.

- *Operational and intelligent transportation system (ITS) strategies*: Improve the operation of the transportation system to make better use of the existing capacity; encourage more efficient driving.

- *Capacity expansion and bottleneck relief*: Expand highway capacity to reduce congestion and to improve the efficiency of travel.

- *Multimodal freight sector strategies*: Promote more efficient freight movement within and across modes.

These strategies to reduce transportation emissions are aimed at affecting three variables: transportation mode (e.g., the type of vehicle or conveyance), travel distance, and efficiency (see box 5.3 for examples). Changing transportation mode is usually described as "alternative

Box 5.3
Examples of Transportation Strategies that Reduce Greenouse Gas Emissions

London's Congestion Charge Zone

The City of London, England, designated a portion of the central and western part of the city as a Congestion Charge Zone (CCZ). Motorists who want to drive into the CCZ during business hours must pay a fee or risk a fine. The intent of the fee is to reduce congestion and generate funds for improvements to the transportation system, especially alternative transportation modes. This is an example of using a pricing strategy to change motorists' behavior.[a]

Portland's Twenty-Minute Complete Neighborhoods

The City of Portland, Oregon, is implementing a land use planning principle that says people should live in neighborhoods that allow them to walk no more than twenty minutes to access basic daily destinations and services such as grocery stores, restaurants and pubs, laundromats, drug stores, parks, and the like. Using this principle to design neighborhoods would require people to drive less and encourage more walking and biking. This has co-benefits of increasing the safety and friendliness of the community as well as the health of the residents.

Pittsburgh's Martin Luther King, Jr. East Busway

The City of Pittsburgh, Pennsylvania, developed a nine-mile bus rapid transit system to link eastern and downtown areas of the city. The roadway is dedicated to buses only (both local and express) thus decreasing travel times during peak commute hours by offering an alternative to crowded roadways. The City is encouraging infill development and redevelopment along the busway to achieve smart growth principles.

San Luis Obispo's South Street Road Diet

The City of San Luis Obispo, California, worked with the state transportation agency to implement a "complete streets" policy on a significant cross-town roadway. South Street was reduced from four lanes to two lanes, bicycle lanes were widened to five feet, transit stops were upgraded, and pedestrian crossings and refuges were added. All of this resulted in no additional vehicle congestion, with improved safety, aesthetics, and community connectedness.

[a] "Case Study: London Congestion Charging," Department of Transport, accessed February 20, 2011, http://www.dft.gov.uk/itstoolkit/CaseStudies/london-congestion-charging.htm.

transportation," and usually involves strategies encouraging community members to change their mode of travel by shifting to walking, bicycling, or transit instead of the private vehicle. Changing travel distance often falls under the concept of smart growth, which recognizes that the distribution of land uses influences travel behavior. Long-term land use planning can aim to shorten the distance between residential areas and common destinations, for example. This reduces the number of miles traveled by vehicle and makes alternative transportation more feasible. Changing efficiency includes a move to more fuel-efficient vehicles (e.g., hybrid or scooters), a change to alternative fuels such as biodiesel, or an increase in the average number of passengers in a vehicle (carpooling). The various ways to address transportation emissions can be taken together to guide urban design principles for accessible services and streets that accommodate all forms of transportation and all members of society, referred to as "complete streets."[9]

Many strategies that reduce transportation-related emissions also provide co-benefits of improvements to air quality and human health and safety. Many lower the costs of transportation since transit, bicycling, and walking are less expensive than driving. In addition, strategies that improve safety of pedestrian travel such as vegetated medians also promote carbon sequestration, improve stormwater management, and enhance overall aesthetics. Land use strategies that place residents and services (schools, employment, grocery, etc.) in close proximity have the potential to promote community cohesion and quality of life.

Shifting to Alternative Transportation Modes
Because the choice to replace a vehicular trip with walking, biking, or public transit is voluntary, education, outreach, and other programs to encourage behavior change are often critical to emissions reduction. In parallel with improved availability of travel options, many cities conduct extensive outreach and incentive programs. These strategies can include bicycle safety education programs, the provision of bicycle and transit maps, and discount transit passes. Another way to influence travel mode is by working with employers to develop incentives for employees who choose to commute using an alternative to a private vehicle.

Encouraging a shift to alternative travel modes requires that walking, bicycling, public transit, or other travel options are convenient and accessible to all community members. Evaluating the opportunity for and constraints to alternative travel modes as well as the factors that influence community willingness will contribute to development of effective policies.

Alternative transportation strategies yield emissions reductions only when they replace a trip that would have yielded higher emissions. Therefore, the types of trips that can be replaced dictate which types of alternative transportation options to emphasize. Walking and biking strategies are most effective for reducing VMT from the portion of vehicular trips that begin and end within the community. Walking and bicycling reduction strategies may include an expansion or improvement of infrastructure (bike routes and sidewalks) to assure that walking and biking can be safely enjoyed by all community members. The emissions associated with trips that originate or end outside the community are best addressed through public transit or ride sharing programs in the short term and by land use that reduces the need for longer trips in the long term. The choice of where to focus efforts can be made through careful evaluation of existing alternative transportation networks and by soliciting community input regarding needed improvements and identification of barriers (e.g., perceptions of danger, convenience, and weather patterns). Bus or train travel can be encouraged through expanded routes, hours of operation, or stops.

In some cases, it may be necessary to combine strategies because emissions reductions cannot be separated. For example, if a community increases the number of bike paths, improves lighting, and adds bicycle storage, it is difficult to assign particular reductions to any one of these strategies individually. In this case, an overall increase in bicycle mode share could be assumed. However, assumptions about eventual shifts in the mode share of bicycling must be supported through demonstration that the community may be able to achieve this level of ridership. A more detailed manner of quantifying transportation shifts can be made by assuming that a portion of the community will increase their reliance on alternative travel options (this can be done separately for walking, biking, and public transit). The next question is to decide what portion of the population will change their travel behavior and what portion of their vehicle trips will be replaced. For example, improved bicycle infrastructure with accompanying education and incentive programs may result in 5% more community members riding their bike. Of these 5%, it could be assumed that they reduce their average daily

Box 5.4
Smart Growth Principles

- Create range of housing opportunities and choices
- Create walkable neighborhoods
- Encourage community and stakeholder collaboration
- Foster distinctive, attractive communities with a strong sense of place
- Make development decisions predictable, fair, and cost-effective
- Mix land uses
- Preserve open space, farmland, natural beauty, and critical environmental areas
- Provide a variety of transportation choices
- Strengthen and direct development toward existing communities
- Take advantage of compact building design

Source: http://www.smartgrowth.org/about/default.asp.

VMT by 50%. In addition, an average fuel efficiency must be identified that provides a GHG per VMT constant. A GHG reduction would then be calculated via the following equation:

$$\text{Estimated GHG reduction} = [\text{Community population} \times 5\%] \times [\text{average daily VMT per capita} \times 50\%] \times \text{GHG/VMT}$$

Growing Smarter to Reduce Travel Distance

Reduced distance between residential areas, employment centers, and services such as grocery stores not only shortens the distance traveled by car but also makes alternative travel modes more convenient. Specific populations and those with the longest commutes should be targeted. Low-income residents may be forced to commute longer distances due to the availability of less expensive housing in outlying areas. An inclusive housing policy in areas closer to jobs and amenities may help to reduce daily miles traveled (often refered to as jobs–housing balance). Reduced travel distances can be encouraged by altering land use policy to encourage smart growth and by providing incentives to developers for infill development or mixed use (box 5.4).

The emissions consequences from mixed use and infill development are difficult to quantify. These changes in land use pattern will alter VMT and will also improve the feasibility of alternative transportation options. Quantifying these changes can be completed as a rough estimate through a reduction in per capita VMT from the business-as-usual forecast in the emissions inventory, but there have been more detailed strategies proposed.[10]

Increasing Travel Efficiency
Encouraging community members to carpool or utilize higher-efficiency vehicles such as hybrids, alternative-fuel vehicles, or high-mileage vehicles (e.g., scooters) can be achieved by making these options more convenient than the alternative. Designated lanes or parking for carpool, hybrid, and high-mileage vehicles provide incentives for use, and increased parking or driving fees (such as congestion pricing) are disincentives for driving.

To justify estimating emissions reduction it can be helpful to assess the effectiveness of reduction programs in other communities. The results from other communities should be adjusted based on local characteristics, but carpooling, parking availability and fees, and congestion management are all strategies that predate climate planning.

Energy Efficiency

GHG emissions are produced in the generation of electricity and through the use of other fuels such as natural gas or propane. These energy sources are used primarily in buildings for electricity, heating, and cooling. Measures that improve energy efficiency reduce GHG emissions. Greater efficiency can be achieved in a variety of ways from energy efficient appliances and fixtures, building materials such as insulation or windows, to solar orientation and use of trees for shade (box 5.5). There is a well-established knowledge base for improved energy efficiency in buildings broadly referred to as green building. Energy efficiency, which includes reduced heating, cooling, and water demand, can be achieved through a variety of complementary strategies. There are many resources from which to draw strategies, including the U.S. EPA,[11] U.S. Green Building Council,[12] and Build It Green.[13]

In addition to building design, energy use in buildings is associated with the behaviors of inhabitants. The choice of indoor temperature and the act of turning off lights and other energy-using appliances, including computers when they are not in use, are examples of how

Box 5.5

Example of Innovative Energy Efficiency Strategy to Reduce Emissions

City of Burlington's (Vermont) Minimum Rental Housing Time of Sale Energy Efficiency Standards Ordinance

According to the city:

> The purpose of the ordinance is to promote the wise and efficient use of energy through cost-effective minimum energy efficiency standards for rental dwellings where physically possible. The ordinance is applied upon transfer of a rental property at the time of sale. The seller and the buyer negotiate who is responsible for compliance. Through the program administrator, technical assistance and coordination with all available energy programs are available to help property owners meet the requirements. In addition, some buildings offer substantial energy savings if work is done beyond the minimum ordinance requirements. Optional technical assistance, project management incentives and financing packages are available to help property owners take advantage of these additional savings.

Source: "Energy Efficiency Codes and Ordinances," Burlington Electric Company, accessed September 10, 2010, https://www.burlingtonelectric.com/page.php?pid=37&name=ee_codes.

inhabitant behavior influences energy demand. Community members may not be aware of the energy demand resulting from various choices. The first step in changing energy use choices is increasing awareness through extensive outreach and education. Altering the pricing structure of energy can also yield changes in behavior, but effectiveness of this measure will likely be higher when paired with outreach.

Programs that target behavior change are often implemented in combination with other strategies such as incentive programs or pricing adjustments. If a program that seeks to alter user behavior is used to support another strategy, there should not be a separate estimate of emissions reduction. Instead the outreach should be viewed as part of the implementation plan for the strategy it supports.

The other area in which energy efficiency can be improved is in governmental services and infrastructure, such as the treatment and conveyance of water and the regulation of streets. The energy required to treat and deliver water throughout a community can be reduced in two ways: improve the energy efficiency of pumps and treatment plant operations, and reduce demand for water, which lowers the volume of water requiring treatment and delivery. Streetlights and traffic signals require a considerable amount of energy in dense urban settings. Technology is currently available to reduce the energy required to power these lights by over 40%.[14]

Energy efficiency strategies have many co-benefits. Retrofit of existing buildings and construction of new energy efficient buildings contribute to a community's resilience in the face of climate impacts. For example, green roofs have the potential to sequester carbon, reduce energy use, and provide protection to inhabitants facing heat-related climate impacts. In addition, a local requirement for energy efficient construction creates a demand for specific construction expertise that has the potential to promote an area of economic growth. Finally, reduced energy demand lowers the monthly costs for residents, making housing more affordable.[15]

Three areas in which to target energy efficiency strategies, existing buildings, new structures, and water treatment and delivery, are discussed in the following sections.

Existing Buildings

Community-wide improvement in energy efficiency must address the structures already in place. In many cases, considerable energy savings can be achieved through retrofitting existing buildings, particularly those that predate modern building codes. The efficiency of these structures can be upgraded in a variety of ways. Buyers of existing buildings or homes can be required to upgrade fixtures such as light bulbs or appliances at the point of sale. Rebate or micro-loan programs are also commonly used to offset the upfront cost for more expensive building retrofits such as insulation or windows.

Many of these strategies not only reduce energy needs but also remove GHGs from the atmosphere. For example, installation of a green roof can improve insulation and natural shading as well as introduce plants that can sequester carbon. Similarly, planting street trees along the sides of buildings with the highest solar exposure reduces energy demand by regulating indoor air temperature, improves the quality of the streetscape, and sequesters carbon.

If a strategy supports or requires retrofitting a particular aspect of building operation, the required information for quantifying the strategy is best obtained from the manufacturer of the item (e.g. appliances, windows, etc.) or from the agency that regulates appliance efficiency (e.g., the Energy Star Program[16] or the Department of Energy Home Energy Saver[17]). The other needed information is an estimate of existing energy use (from the emissions inventory) and participation rates.

New Structures
New building requirements are more easily implemented because the energy-saving strategies can be included in the design and budgeting of a project and can be a requirement for permitting. The manner in which new development is regulated varies, meaning the opportunities for promoting or mandating green building will be similarly varied. Specific green building policy can take actions, such as requiring energy efficiency as a condition of permit approval, which is often referred to as a green building ordinance. Such requirements can also be part of a larger green building program where minimum building standards are set and incentives such as expedited processing are provided for more advanced green building techniques.

Water Treatment and Delivery
Reducing the energy required to treat and convey water and wastewater can be achieved through upgrading the pumps and other equipment required for treatment and delivery or reducing community water demand. Decreasing water demand can be achieved by upgrading water fixtures (such as toilets, sinks, and showers). It can also be achieved through increased participation of households in rainwater capture or graywater systems or citywide water recycling. These strategies can be implemented to serve nonpotable uses such as irrigation of yards and landscaped areas.

Water use can also be limited through landscaping and yard vegetation choices. Outdoor water use is one of the largest consumers of potable water in the United States. Vegetation choices that require less water can result in substantial reductions in water use for residential yards and irrigated park areas. Communities can promote these changes through educational materials such as planting guides, incentives such as cash-for-grass programs that compensate the removal of residential lawns,[18] or mandates such as vegetation and irrigation requirements on building permits. Quantification of these measures draws heavily on data in the emissions inventory, including a conversion factor of CO_2e per gallon of treated water and the average water use per household. Using these constants, the estimated gallons of water saved can be converted to CO_2e based on the anticipated effectiveness and participation in a program. Energy use by existing pumps can be directly measured, and new pumps are rated, so assessing the efficiency of water conveyance pumps and lifts is relatively straightforward.

Renewable Energy

Renewable energy, such as solar, wind, or biomass, provides electricity and heat without the same level of GHG emissions associated with traditional energy sources. The addition of local renewable energy generation lowers GHG intensity (GHG per unit energy). The largest deterrent for renewable energy is the initial cost of installation. The most appropriate type of renewable energy will vary regionally based on factors such as solar exposure, available surface or land area, wind speed, biomass sources, coastal conditions, geothermal resources, and social acceptance of the proposed technology. Once a renewable energy technology or suite of technologies has been selected, implementation requires a series of actions, including a funding mechanism, the choice between distributed and centralized generation, and the phasing of implementation. Funding renewable energy has been an area of considerable innovation and creativity (box 5.6). Traditional funding mechanisms include allocation of local funds, external investment by the private sector, or grant dollars procured from an outside entity. In addition to these funding sources there are increasingly creative means of funding renewable energy, including providing investment opportunities for local residents, developing a micro grant or loan program, and funding renewable energy through impact fees for environmentally damaging actions.

Quantifying reductions in GHG emissions from renewable energy requires that assumptions be made about the efficiency of energy technology (e.g., wind and solar energy capture rates), the local availability of these sources, and the potential locations for installation. National maps of solar and wind potential are now available and, in many regions, maps have been generated that have higher resolution. Including renewable energy in a CAP has the potential to complicate the quantifi-

Box 5.6
Examples of Innovative Renewable Energy Strategy
to Reduce Emissions

Salt Lake City's (Utah) Solar Power Purchase Agreement

Salt Lake City is covering the entire 600,000-square-foot Salt Palace Convention Center building with a $10 million solar panel installation.[a] It will produce about 2.6 megawatts of electricity, which is a quarter of the building's annual electricity demand. The project is an example of a power purchase agreement (PPA). In a PPA a private company builds and operates the solar array and charges the City through utility billing. The City bears no up-front costs and pays for electricity it would use anyway. The Salt Palace project is also partially funded by federal grants and tax credits, and the City will have an option to eventually purchase the solar array at a steep discount.

Marin County's (California) Marin Clean Energy Program

In Marin County, communities had worked together to establish Marin Clean Energy (MCE), an example of community choice aggregation (CCA). CCA is the aggregation of electricity demand among various community users (residents, businesses, etc.) to facilitate the purchase of electrical energy, especially from renewable energy sources. The MCE program automatically subscribes electricity customers in the participating communities to the program. MCE then purchases renewable energy in cooperation with the franchised commercial energy provider. In 2010, MCE was supplying power that was 25 percent from renewables (double the commercial energy provider) and expects to be at 50 percent by 2015.

[a] John Daley, "Salt Palace to House Largest Solar-Power Installation in Country," KSL.com, September 1, 2010. http://www.ksl.com/?nid=148&sid=12246453.

cation of energy efficiency measures. Because renewable energy changes the energy intensity (GHG/kWh), it also influences the reductions experienced as part of improved efficiency. As the percentage of the electricity supplied from renewable sources increases, the GHG reduction from efficiency measures decreases. This is a concern for communities with aggressive renewable energy goals. It can also influence phasing of energy strategies where efficiency makes most sense as a short-term goal.

One of the important co-benefits of renewable energy programs is that they foster local economic growth by employment of the workforce needed to install the systems and, if materials are manufactured locally, the employees of the manufacturer. In addition, renewable energy increases human and ecosystem health due to removal of air pollution associated with energy generation from fossil fuels.

Carbon Sequestration

In addition to trying to reduce emissions, climate strategies may take the approach of capturing some of the carbon and sequestering it in terrestrial vegetation (e.g., trees) and soil. Terrestrial vegetation, such as forests, sequesters carbon through increased volume of woody mass. Trees can be used as a shade crop in agricultural practices, as street trees in cities, and for larger-scale reforestation projects. Particularly in urban areas, trees provide shade for structures that improve energy efficiency and the pedestrian environment, making alternative transportation more appealing. The type of vegetation should be considered carefully to assure consistency with the local climate and soil conditions, as well as the intended role of the vegetation in addition to carbon capture.

Soil carbon sequestration refers to the organic content of soils such as leaf litter and other biomass. Strategies to increase soil carbon content directly address widespread soil degradation.[19] To sequester carbon, the soil carbon must be stored long term and not released back into the atmosphere.[20] Successful implementation of soil sequestration strategies also improves soil and agricultural productivity. A variety of methods can be used to increase soil carbon levels that should be chosen based on local conditions, including no-till or conservation tillage farming practices, use of cover crops, managing the nutrient input to soils, agroforestry, woodland regeneration, crop rotation, and improved grazing practices.[21]

Quantification of GHG reductions due to sequestration relies on improvements in vegetative uptake or change in conditions from the baseline established in the emissions inventory. Newly planted vegetation such as street trees or wind breaks and newly introduced soil management practices are quantified, whereas contributions from existing green spaces that predate the climate planning effort are not because they are assumed to be included in the baseline. The estimated sequestration possible for vegetation such as street trees varies by climate region, tree species, and age. Many state forestry departments and universities provide lists that estimate sequestration potential that is regionally accurate. Similarly, soil carbon content varies by climate, soil type, and land activity.

Agriculture Management

Agriculture, which accounts for 7% of U.S. GHG emissions, has been identified as an area with significant emissions reduction potential through carbon sequestration and alternative management practices.[22] Thus far, climate planning has been largely focused on cities, but the more recent emergence of regional efforts and county plans has required agriculture to be addressed directly.

Including rural areas in climate planning will require the measurement of annual variability in emissions associated with agriculture. The fluctuations in agricultural emissions results from variation in climate, soil type, and agricultural practice. In addition, the longevity and permanence of sequestration efforts can vary. Carbon sequestration in agricultural lands is effective for a specific duration of time (e.g., 15–30 years). Shifts in management practices can reverse the benefits resulting from a reduction strategy. Many argue that this results in agriculture being best pursued in the near term providing time for more expensive or time-intensive strategies such as large-scale renewable energy or land use change to be implemented.

Agricultural practice is vulnerable to potential shifts in temperature and precipitation that are projected to result from climate change. It is possible to devise reduction strategies that have the co-benefit of bolstering adaptive capacity, but not all strategies meet both reduction and adaptation goals. For example, increased food production intended to bolster local food security can be achieved through converting additional land to agriculture and/or increasing the application of fertilizers. Both of these actions have the potential to increase emissions. Conversely, some reduction strategies have the potential to reduce production, placing it in conflict with adaptation needs. As a result, agricultural reduction strategies must be carefully identified and constructed to assure the best balance between reduction, adaptation, and local needs such as food supply, ecosystem protection, and local employment. Reduction strategy development must be conducted in close collaboration with local agricultural communities to assure feasibility and regionally appropriate strategies.

A wide variety of strategies can reduce agricultural GHG emissions (box 5.7). The choice of action will depend on the factors such as the local environmental condition, type of agriculture, current management practices, soil properties, local climate, economics, and local work

Box 5.7
Carbon Sequestration and Agriculture Strategy Summary

Urban land: tree planting, waste management, wood product management

Cropland: reduced tillage, rotations and cover crops, nutrient management, erosion control and irrigation management, organic farming

Agroforestry: better management of trees on croplands and conversion from unproductive cropland and grasslands

Grazing land: improved grazing and rotation practices, woody plant, and fire management

Forestland: forest regeneration, fertilization, choice of species, reduced forest degradation

Restoring degraded land: change to crop, grass, or forestland

Rice paddies: irrigation, fertilizer management, and plant residue management

Grassland: conversion of cropland to grassland

Livestock: more easily digested feed, more monogastric animals, herd health programs, or anaerobic manure digesters producing biogas

Source: Adapted from Intergovernmental Panel on Climate Change, Land Use, Land Use Change and Forestry, Summary for Policymakers (IPCC, 2000); Food and Agriculture Organization of the United Nations, Soil Carbon Sequestration for Improved Land Management (Rome: Author, 2001); N. V. Nguyen, *Global Climate Changes and Rice Food Security* (Rome: International Rice Commission, Food and Agriculture Organization of the United Nations, 2004); P. Smith, D. Martino, Z. Cai, D. Gwary, H. Janzen, P. Kumar, B. McCarl, et al. "Policy and Technological Constraints to Implementation of Greenhouse Gas Mitigation Options in Agriculture," *Agriculture, Ecosystems and Environment*, 118 (2007): 6–28; H. Steinfeld, P. Gerber, T. Wassenaar, V. Castel, M. Rosales, and C. de Haan, *Livestock's Long Shadow Environmental Issues and Options* (Rome: Food and Agriculture Organization of the United Nations, 2006).

force. Reduction strategies can be broken down into a set of broad categories: carbon sequestration (discussed in the prior section), livestock management, and rice paddy and wetland strategies (not addressed in this book).

Livestock Reduction Strategies

Livestock has climate and other environmental impacts associated with soil degradation, methane output, biodiversity loss, water use, and land use change. The strategies discussed earlier as part of the sequestration section also apply to grazing lands. This section focuses on two sources of emissions associated directly with animals: ruminant digestion and manure management. Ruminants (cows, goats, sheep, llamas, etc.) release methane, which has twenty-one times the global warming potential of CO_2, as part of the digestive process. Methane emissions are higher when an animal's diet is poor.[23] One manner in which to curb these emissions is through improved nutrition. There are feedstuffs that have increased digestibility that can take the form of feed additives. Another strategy is a move to more efficient animals that are monogastric, such as poultry.[24] The expense of changing animal feed is partially offset by findings that animals grow larger and milk production increases with the more easily digested feed. Other strategies that reduce methane production are herd health programs.

The other source of GHGs associated with livestock animals is derived from manure, which also results in methane production. Similar to the direct emissions, a shift in livestock feed can limit some of the methane production. A low carbon-to-nitrogen ratio results in increased emissions from manure. If the collected manure can be stored at a warmer temperature, or outdoors in temperate climates, emissions will be lower. The manure can also be handled in a digester, a closed vessel with controlled conditions. Technology already exists not only to reduce the emissions but to generate energy from the biogas produced in the digester.[25]

Quantification of agricultural measures should be tied directly to assumptions in the emissions inventory. If the inventory includes agriculture, an emissions rate per head of livestock for manure disposal will have been established. Reductions based on changes in feed, land management, or manure handling should be calculated based on improvements from baseline. In the case where manure is used to generate methane, the production of energy can be gathered from the information provided by the manufacturer or supplier of the digester.

Industrial Facilities and Operations

Industrial sector emissions present a special challenge for communities. Since most aspects of industrial operations are regulated at the regional, state, and federal level, local governments have little ability to mandate industrial changes. The approach with the industrial sector should focus on outreach and partnership. Not only should awareness of climate change and reduction strategy development process be promoted, but the concerns and goals of the industrial sector should be solicited and considered in strategy development. Reduction strategies that focus specifically on this sector should be developed in a manner that seeks to assure that long-term emissions reduction goals are compatible with long-term local economic viability. Many communities institute programs to rate and publicize local green business. Programs such as these seek to provide visibility for climate-friendly businesses. Energy efficiency improvement can also be mandated by placing requirements on business licenses. These can include measures from those described in the early sections for energy efficiency in buildings to measures to incentivize climate-friendly employee commuting behavior.

Many GHG reductions in the industrial sector can be achieved through energy efficiency strategies included in the building and renewable energy sections. Industrial structures can be upgraded for efficiency, and the large roof surfaces of many industrial structures are ideal for installation of photovoltaic panels or a green roof. Emissions reductions can also be achieved through changes in operational procedures and in the relationship between industries in the same community. Operations can refer to a variety of factors from the efficiency of machinery to the vehicles used on site. Strategies regarding the relationship between industrial entities are captured in the principles that govern eco-industrial parks. An eco-industrial park is an industrial complex that seeks to collectively manage resource use, energy, and material flows for enhanced efficiency and improved environmental performance. For example, this would include pairing companies where the waste product from one industrial process can serve as the input for another.

Quantification of industrial measures draws on those methods described for energy efficiency in buildings and renewable energy. In the case of eco-industrial parks, the reductions can be determined by calculating the reduced miles traveled by trucks hauling waste off site and input materials on site. The local values for GHG per heavy-duty-truck mile can be obtained from the emissions inventory. Additional reductions can be calculated on a case-by-case basis using the baseline assumptions defined in the inventory.

Waste

Waste deals with the treatment or disposal of postconsumer solid waste and waste resulting from the treatment of wastewater that generates methane during decomposition. Reducing emissions from waste treat-

ment can be achieved in two ways: reduce the amount of waste produced and reduce the emissions associated with waste disposal. This
twofold approach is necessary because waste disposed in a landfill will
emit methane for decades following the initial disposal of the waste. As
a result, reduction of waste will not lower landfill emissions in the near
term, although some methods for assessing the emissions reduction from
landfill diversion account for this by annualizing emissions. What will
decrease immediately with reduced waste production are GHG emissions associated with collection, delivery, and handling of waste.

Co-benefits of waste strategies include environmental benefits
from reduced consumption of land for disposal to improved air quality
that will result from reduced collection and delivery vehicle emissions.
Consumption and disposal behaviors also contribute to overall community sustainability.

Waste production

Reduced waste production can occur through both government action
and community behavior change; long-term success will rely on both.
Recycling programs are some of the most well-established means of
solid waste reduction. A city can increase the local diversion rate (percentage of waste stream recycled) by increasing the number of products
that can be recycled, increasing the convenience of recycling through
provision of bins and pickup services, making recycling mandatory or
providing incentives, and conducting outreach to increase the participation rate of the community. An emerging area of emphasis in recycling
is disposal of e-waste such as computers. Programs can be developed to
safely disassemble e-waste to recover resources. Waste reduction can also
be achieved through strategies to reduce packaging and other materials
that accompany products.

Organic waste from food or outdoor vegetation is another opportunity for waste diversion that directly addresses the organic matter that
generates methane. Strategies for addressing organic matter include
composting and converting vegetative material to mulch. These strategies can be implemented on a city scale or on an individual scale depending on housing type. A city-scale program will require a facility
designed to accept the waste, a means for waste to reach the facility, and
a program to encourage participation. The challenge of individual composting or yard waste programs is participation at a level that impacts
GHG emissions. Both the City of Seattle[26] and the City of San Fran-

cisco[27] have instituted residential curbside composting programs. In both cases the resulting compost is used in local parks and agricultural areas.

Waste Disposal

Landfills and wastewater sludge both generate methane that can be captured and converted to electricity. This requires that waste be covered and the landfill or collection point be retrofit for technology to allow methane capture. This is an immediate reduction in methane generation associated with waste disposed of in the past. Other disposal methods pose their own sets of challenges. Incineration or waste-to-energy plants primarily emit carbon dioxide instead of methane. While carbon dioxide is a less powerful GHG than methane, waste-to-energy plants also emit a variety of other air pollutants that contribute to acid rain and may pose a threat to human health. The benefit of these plants is that they do not consume the land area of a landfill.

Green Living

Green living refers to reduction strategies aimed at daily behaviors in the home and workplace that may not be covered by other sectors. These may include strategies to motivate people to eat more locally and sustainably, grow their own food, or change their purchasing to more environmentally friendly products (Box 5.8). These reduction strategies are difficult to quantify and are usually favored more for their co-benefits than for their potential GHG emissions reductions. They are more likely to be found in sustainability or green plans.

Carbon Offsets

Offset or carbon offset programs are designed to deal with difficult-to-reduce GHG emissions occurring in one sector or community by taking action to lower emissions elsewhere. For example, if a city cannot control GHG emissions from its electric utility provider it may choose to offset those emissions by planting trees in a forest in the region (to act as a carbon sink). Offsets can be managed through compensatory or reciprocal strategies or they can be structured financially through purchase arrangements. The purchase of offsets is accomplished by setting up a financial system where individuals, businesses, or communities that create GHG emissions offset those emissions by purchasing credits or paying into a fund. The credits or fund are then used to finance other

Box 5.8
Example of a Green Living Program from Davis, California

Try the Cool Davis Low Impact Challenge—Take One Simple Step to Reduce Your Carbon Footprint Each Day over a Week

All Davis households are encouraged to participate in the Cool Davis Low Impact Challenge. The Challenge is aimed at getting the citizens of Davis to reduce their carbon impact by taking one or two different simple, concrete, positive steps each day for five days to reduce their carbon emissions and ease into a more sustainable, greener way of life.

Day 1—Cut the Trash!

On this first day of the challenge, we invite you to generate less throw-away trash by trying one or more of these steps:

1. Use cloth bags (not plastic or paper) to bring home groceries or other purchases.
2. Give up bottled water; tap water has a lower carbon footprint, and will save you money.
3. Carry your own cup or thermos to use throughout the day; avoid disposable cups.
4. Recycle any and all plastic, paper, and metal you use today.
5. Sign up for a city class on composting.

Day 2—Bon Appetit!

Today we invite you to think about the food you eat and its relationship to your health. Try one or more of these steps:

1. Try a meatless day and be kind to your heart.
2. Buy groceries from the Davis Farmers Market, Davis Food Coop, or grocery stores that offer locally grown food and lower your carbon "food-print."
3. Eat organically grown foods and keep both you and our planet healthier.

Day 3—Get Moving!

On this 3rd day of the Challenge, we ask that you use your travel time to get healthier, save money, and enjoy the scenery, too. Try one or more of the following:

1. Got errands in town? Use a bike. Dress in layers to keep warm in winter and cooler in summer.
2. Plan your day so that you have time to take a walk.
3. Just for fun, explore Davis by following the green bike loop around town or venture further using county roads for great views.
4. If you must travel to work or school, use Unitrans, Yolobus, or the Capital Corridor Train.
5. If a car is your only choice, check the air pressure on your automobile tires. Your pocketbook swells as your mpg increases.

Day 4—Save $$$ Energetically!

Today's Challenge is to Save Energy. Try one or more of the following ways to use less energy.

1. If it is cold outside, put on layers to keep you warm when you lower the thermostat. Shed clothes and turn on fans to keep cooler in summer.

2. Wash your clothes in cold water. With modern soaps, they get just as clean.
3. Hang your clothes up to dry and avoid the energy-guzzling dryer.
4. Turn off any electrical devices not in immediate use.
5. If you haven't done so already, install energy efficient compact fluorescent (CFL) bulbs.

Day 5—Water Ways and Days

On this last day of our Challenge, we invite you to use less water. With water becoming more precious and rates rising, everybody wins by conserving. Try one or more of these steps:
1. Limit your shower to 4–5 minutes, and daydream while you dry yourself off.
2. Run the dishwasher only when full and use the no-heat wash, rinse and dry cycles. It is simple, easy and costs nothing to implement this plan.
3. Wear your outer clothing more than once before putting it in the laundry. Less washing saves both water and energy, plus the clothes last longer.

Source: Cool Davis Foundation, accessed March 1, 2011, http://cooldavis.org/Cool%20Davis%20Low%20Impact%20Challenge.htm.

emissions-reduction strategies. Offset programs are very popular at the international level, and many industrialized nations are participating in programs that offset some of their emissions by investing in energy efficiency and renewable energy programs in developing countries; thus the programs provide economic development and social justice benefits as well.

Several important issues must be addressed when considering off-set programs. First, if the offset program works through purchases then a mechanism must be in place to manage and track financial transactions. This could be done by the local government, a local nonprofit, or one of the established international programs such as TerraPass.[28] Second, the location of the offset projects is important. Most communities have developed policies that require offset funds to be spent in their community. This promotes local investment and benefits but may limit the number or quality of opportunities. The last issue is a set of related questions about who pays, how much, for what strategies, and whether it is voluntary or not.

Local examples of offset programs are few. The City of San Francisco, California, set up the San Francisco Carbon Fund to invest in local sustainability projects. Funds come from a 13% surcharge on all city employee air travel and carbon offset kiosks in San Francisco International Airport (Climate Passport Program). The City of Denver, Colorado, had a similar program, but it was suspended because the city was unable to find a vendor to operate the service. Evanston, Illinois (Evanston Climate Action Fund), and Chicago, Illinois (Chicago Offset Fund), are also in the early phases of development of offset programs. Depending the program, GHG reduction as a result of carbon offset can be quantified on a dollar basis (GHG/$) or directly if funds are used for projects such as reforestation where GHG reduction can be estimated based on setting, number, type, and age of trees planted.

Chapter Resources

Quantification of Reduction Measures

California Air Pollution Control Officers Association (CAPCOA), *Quantifying Greenhouse Gas Emissions Measures: A Resource for Local Government to Assess Emission Reductions from Greenhouse Gas Mitigation Measures* (Sacramento, CA: Author, August 2010). http://www.capcoa.org/. The guide provides detailed quantification methods that can be used throughout California and adapted for use outside of the state. It contains a series of fact sheets on particular types of reduction strategies and accompanying guides on how to use the fact sheets.

Climate and Air Pollution Planning Assistant (CAPPA) Software. This is an Excel-based decision support tool designed to help U.S. local governments explore, identify, and analyze potential climate and air pollution emissions reduction opportunities. Developed and maintained by ICLEI. http://www.icleiusa.org/action-center/tools/cappa-decision-support-tool/

Reduction Strategies

Cambridge Systematics, *Moving Cooler: An Analysis of Transportation Strategies for Reducing Greenhouse Gas Emissions* (Washington, DC: Urban Land Institute, 2009). http://www.movingcooler.info/. The main document provides good explanation and insight, and the appendices, which are free to download, provide detailed reasoning for GHG emissions reduction calculations. The constants and assumptions presented in the appendices can be adjusted to reflect local conditions.

Reid Ewing, Keith Bartholomew, Steve Winkelman, Jerry Walters, and Don Chen, *Growing Cooler: The Evidence on Urban Development and Climate Change* (Washington, DC: Urban Land Institute, 2008). http://www.smartgrowthamerica.org/gcindex.html. The report explains the relationship among land use, transportation, and climate change. It explains smart growth strategies, including infill development, mixed use, and sprawl-reducing strategies, and includes examples.

National Complete Streets Coalition. http://www.completestreets.org/. The Coalition's website has numerous resources to help communities plan complete streets, including guidance documents, model policies, examples, and education materials. They state the following: "Complete streets are designed and operated to enable safe access for all users. Pedestrians, bicyclists, motorists and transit riders of all ages and abilities must be able to safely move along and across a complete street."

U.S. Department of Energy, *Solar Powering Your Community: A Guide for Local Governments* (Washington, DC, 2011). http://solaramericacommunities.energy.gov/resources/guide_for_local_governments/. "The U.S. Department of Energy developed this comprehensive resource to assist local governments and stakeholders in building sustainable local solar markets. . . . This updated edition also contains the most recent lessons and successes from the original 25 Solar America Cities and other communities promoting solar energy. The guide introduces a range of policy and program options that have been successfully field tested in cities and counties around the country."

U.S. Department of Energy, Office of Energy Efficiency and Renewable Energy (EERE). http://www.eere.energy.gov/. The EERE website provides information on energy efficiency for homes, buildings, vehicles, industry, and government. It also addresses renewable energy such as solar, wind, water, biomass, geothermal, and hydrogen and fuel cells.

U.S. Environmental Protection Agency, *Sustainable Design and Green Building Toolkit for Local Governments*, EPA 904B10001 (Washington, DC, June 2010). http://www.epa.gov/region4/recycle/green-building-toolkit.pdf. "The U.S. Environmental Protection Agency (EPA) developed the Sustainable Design and Green Building Toolkit for Local Governments (Toolkit) in order to assist local governments in identifying and removing barriers to sustainable design and green building within their permitting process. This Toolkit addresses the codes/ordinances that would affect the design, construction, renovation, and operation and maintenance of a building and its immediate site."

John Randolph and Gilbert Masters, *Energy for Sustainability: Technology, Planning, Policy* (Washington, DC: Island Press, 2008). This text provides an easy-to-understand introduction to energy efficiency, renewable energy technology, and associated policy, from national energy markets to local planning strategies. This is a great resource for planners new to the topic, presenting both the history of energy and a look toward the future and emerging technologies. The length can be daunting (816 pages), but the book can serve as an informational reference where you can skip immediately to the sections particularly relevant to a topic of interest.

Chapter 6

⚜

Climate Change Adaptation Strategies

Adaptation strategies prepare a community to be resilient in the face of unavoidable climate change impacts. Climate impacts such as sea level rise, temperature changes including extreme heat events, and change in precipitation patterns can have a variety of secondary impacts on community conditions from human health and safety, to economics, to ecosystem integrity. The challenge of developing effective community adaptation policy is the need to apply the evolving science that describes a global phenomenon at a regional and local level. The inherent difficulty in projecting global climate change impacts is amplified at these levels; currently, regional and local forecasts of the impacts of climate change are considered to be very uncertain.[1] Handling this uncertainty in a policy context requires a combination of flexibility, a willingness to adapt, and careful evaluation of potential climate impacts in the local context.

The evaluation of the threat posed by climate change includes the nature of the projected impact (e.g., rate and magnitude of change), the exposure and sensitivity of the local community and setting to these changes, and the local capacity to adapt.[2] Adaptation strategies seek to reduce vulnerability to projected changes and improve the ability to adjust to unexpected consequences. Strategies can be formulated and prioritized based on locally developed criteria that allow for clear identification of those areas most in need of action. Increasingly, climate adaptation is explicitly included as part of a climate action plan (CAP). Some communities have even chosen to develop a free-standing climate adaptation plan to complement a CAP that focuses solely on greenhouse gas (GHG) reduction. In addition, adaptation strategies can be integrated with existing local plans such as a comprehensive plan or local hazard mitigation plan.

Climate adaptation strategies cover many of the same issues as those that are addressed in natural hazards mitigation, but there are two key differences. First, climate adaptation addresses some climate impacts that are not traditionally defined as hazards such as ecosystem changes. Another difference between climate adaptation and natural hazards mitigation is that history alone is no longer an accurate predictor of future risk and therefore is an inadequate basis for policy development. For example, the projected severity and frequency of natural hazards such as floods can no longer be defined by past occurrence alone (as it has been traditionally defined in hazards planning). Many climate impacts are still episodic events (e.g., flood, fire, drought), but the frequency and severity of the episodes are changing through time. Most communities have strategies in place through the safety element of a comprehensive plan or local hazard mitigation plan that address these episodic events or hazards. In the short term, some climate impacts may be addressed by bolstering existing measures intended to address natural hazards. However, in the long term, these existing strategies may not be adequate to address impacts as they deviate further from historical patterns. In some cases, climate change impacts have the long-term potential to be catastrophic for community health, safety, and economic stability if a community is unprepared. Adaptation strategy development should be seen as an opportunity for communities to position themselves for long-term resilience.

This chapter examines the primary steps (fig. 6.1) in developing strategies for adapting to climate change impacts and begins with identification of the issues that a community should examine as part of the strategy development process.

Issues in Climate Adaptation Planning

Before starting a climate adaptation process, there are several issues that should be considered since they will affect strategy development. These include addressing how the emissions reduction and adaptation components of the plan will relate to each other, how adaptation planning will be coordinated with local hazard mitigation planning, how to deal with uncertainty, and the meaning of the resilience concept for local adaptation. In addition, the White House Council on Environmental Quality,

Figure 6.1   Climate change adaptation strategy development process.

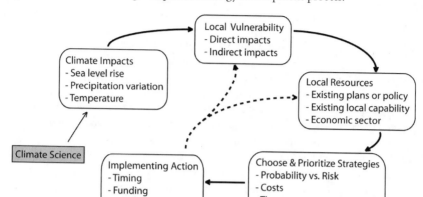

in its report *Progress Report of the Interagency Climate Change Adaptation Task Force: Recommended Actions in Support of a National Climate Change Adaptation Strategy*, provides a set of principles to guide adaptation planning (box 6.1).

Relationship to Emissions Reduction

Emissions reduction and adaptation goals are complementary in many ways (fig. 6.2), but do have the potential to conflict. It is also important to recognize that the considerations which contribute to strategy development differ, even if some of the measures are ultimately similar. A particular adaptation need, such as protection against extreme heat, can be addressed in a variety of ways. GHG reduction should be considered a potential co-benefit for adaptation measures, but this is secondary to the requirement that measures adequately address the scale and severity of the climate impact. For example, tree planting both sequesters carbon and helps alleviate the impacts of extreme heat, while strategies such as cooling centers that offer protection from heat may rely on air conditioning, which can be associated with the release of GHGs due to energy use. The trees address both emissions reduction and adaptation, but they may not offer protection from heat for the most vulnerable populations in a community, making the cooling centers a short-term necessity. There is disagreement among climate policy experts regarding

Box 6.1
Guiding Principles for Adaptation

Adopt integrated approaches: Adaptation should be incorporated into core policies, planning, practices, and programs whenever possible.

Prioritize the most vulnerable: Adaptation plans should prioritize helping people, places and infrastructure that are most vulnerable to climate impacts and be designed and implemented with meaningful involvement from all parts of society.

Use best available science: Adaptation should be grounded in the best available scientific understanding of climate change risks, impacts, and vulnerabilities.

Build strong partnerships: Adaptation requires coordination across multiple sectors and scales and should build on the existing efforts and knowledge of a wide range of public and private stakeholders.

Apply risk management methods and tools: Adaptation planning should incorporate risk management methods and tools to help identify, assess, and prioritize options to reduce vulnerability to potential environmental, social, and economic implications of climate change.

Apply ecosystem based approaches: Adaptation should, where relevant, take into account strategies to increase ecosystem resilience and protect critical ecosystem services on which humans depend to reduce vulnerability of human and natural systems to climate change.

Maximize mutual benefits: Adaptation should, where possible, use strategies that complement or directly support other related climate or environmental initiatives, such as efforts to improve disaster preparedness, promote sustainable resource management, and reduce greenhouse gas emissions, including the development of cost effective technologies.

Continuously evaluate performance: Adaptation plans should include measureable goals and performance metrics to continuously assess whether adaptive actions are achieving desired outcomes.

Source: From The White House Council on Environmental Quality, *Progress Report of the Interagency Climate Change Adaptation Task Force: Recommended Actions in Support of a National Climate Change Adaptation Strategy* (Washington, DC: Author, October 5, 2010), 10.

the relative importance of these goals. Some believe that reduction goals should always be prioritized over adaptation goals,[3] but adaptation has been the focus of increasing attention in many recent guidelines.[4] We firmly believe that the two overarching goals of emissions reduction and adaptation should be treated as equal with the relative priority of emissions reduction versus adaptation made on a strategy-by-strategy basis, with consideration being given to local pressures and needs.

Figure 6.2   An illustration of the overlap between greenhouse gas emissions reduction and climate adaptation measures.

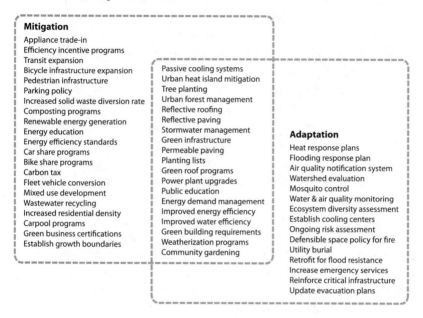

Mitigation

Appliance trade-in
Efficiency incentive programs
Transit expansion
Bicycle infrastructure expansion
Pedestrian infrastructure
Parking policy
Increased solid waste diversion rate
Composting programs
Renewable energy generation
Energy education
Energy efficiency standards
Car share programs
Bike share programs
Carbon tax
Fleet vehicle conversion
Mixed use development
Wastewater recycling
Increased residential density
Carpool programs
Green business certifications
Establish growth boundaries

Passive cooling systems
Urban heat island mitigation
Tree planting
Urban forest management
Reflective roofing
Reflective paving
Stormwater management
Green infrastructure
Permeable paving
Planting lists
Green roof programs
Power plant upgrades
Public education
Energy demand management
Improved energy efficiency
Improved water efficiency
Green building requirements
Weatherization programs
Community gardening

Adaptation

Heat response plans
Flooding response plan
Air quality notification system
Watershed evaluation
Mosquito control
Water & air quality monitoring
Ecosystem diversity assessment
Establish cooling centers
Ongoing risk assessment
Defensible space policy for fire
Utility burial
Retrofit for flood resistance
Increase emergency services
Reinforce critical infrastructure
Update evacuation plans

Relationship to Local Hazard Mitigation Planning

Natural hazards have a much longer history of being addressed through planning than climate change and can provide lessons for climate adaptation. To adequately address climate change, the natural hazards planning process must recognize that historic patterns of natural hazard occurrence alone are no longer appropriate predictors for future occurrences. This change does not invalidate natural hazard planning tools, but it does mean that these tools require adjustment and updating to accommodate the evolving nature of the hazards being addressed.

Communities may engage in some form of local hazard mitigation planning. This could be through safety or hazard mitigation elements of a comprehensive plan, or it may be through preparation of a plan under the federal Disaster Mitigation Act of 2000 (DMA 2000). The DMA 2000 Local Hazard Mitigation Plan (LHMP) program provides communities with a financial incentive, through grant eligibility, to prepare an LHMP. Hazard mitigation is defined as "sustained action taken to reduce or eliminate long-term risk to people and their property from hazards."[5] This should not be confused with "climate mitigation," which refers to the reduction of GHG emissions. Nor should hazard mitigation be confused with other aspects of emergency management such as pre-event preparation or emergency response. Hazard mitigation includes actions such as floodplain regulation, seismic building code enforcement, and vegetation management for wildfire.

The process of hazard mitigation planning is well established and is instructive for informing adaptation planning; the steps and logic are similar. It is based around the core idea of risk assessment, which includes identifying hazards, profiling hazard events, inventorying community assets, and estimating the potential losses from disasters. This risk assessment then informs the development of hazard mitigation strategies for the community.

Since climate change has the potential to alter the type, frequency, and severity of natural hazards, it will affect a community's risk assessment. Any work on adaptation planning should be coordinated through a revised risk assessment that accounts for the impact of climate change on natural hazards in the community. This revised risk assessment would then inform adaptation planning, local hazard mitigation planning, and community land use planning. The strategies in these various planning documents can then be coordinated or integrated to comprehensively address risks that currently exist and future risks that will be influenced by climate change.

Dealing with Uncertainty

Planning for adaptation relies more directly on climate science than planning for emissions reduction. The climate impacts being projected by scientists often have direct, concrete costs to a community. Adaptation strategies respond to these impacts, many of which have the potential to require large capital investment, in contrast to emissions reduction strategies, which seek to lower global concentrations. This tangible risk demands explicit accounting of uncertainty in strategy development. It can be difficult for local communities to choose to make such investments to prevent or reduce vulnerability to potential climate change impacts that have high levels of uncertainty or a low probability of occurring. Part of the strategy development process requires communities to evaluate potential impacts in terms of level of risk (to property, safety, economy, health, etc.). There are impacts that despite having a low probability and high levels of uncertainty will require action due to the scale, severity, and irreversibility of the losses that would result.

The Concept of Resilience

Climate adaptation refers broadly to measures that increase the ability of a community to withstand climate impacts. This is often referred to as adaptive capacity or resilience. The term *resilience* can be used in an engineering sense to describe structural performance or as a system property. In the case of local planning, the system is the community, defined by the interacting elements of the biophysical setting, built environment, and sociopolitical conditions. The resilience concept can be most clearly understood by considering two broad forms:[6]

1. The direct strength of structures or institutions when placed under pressure
2. The ability of systems to absorb the impact of disruptive events without fundamental changes in function or structure.

Adaptation planning strives for resilience (box 6.2). The engineering or structural resilience can be seen as a component of system resilience. Resilience allows for strategies to be viewed in a larger context and can be freeing for the process of policy development. For example, the strengthening of levees in flood-prone areas such as New Orleans improves structural resilience in the face of sea level rise and increased hurricane intensity. But this measure alone does not address citywide or systematic resilience. Addressing adaptation on a local scale may result in policy that does not act directly on a projected impact. Strategies aimed at system resilience may include those that improve the flexibility of the economic sector so that it may adjust more quickly to changes brought about by climate change, or policy measures may aim to improve conditions for populations that are disproportionately vulnerable to impacts.

The Adaptation Strategy Development Process

The adaptation strategy development process consists of five steps as shown in figure 6.1:

Box 6.2
Attributes of Resilient Communities

- Diversification of livelihood activities, assets, and financial resources particularly into activities that have low levels of sensitivity to climatic variability or extreme events
- Mobility and communication, particularly the ability of goods, people, information, and services to flow between regions in ways that enable local populations to access markets, assets, the media, and other resources beyond the likely impacts of specific climatic events
- Ecosystem maintenance, particularly maintenance of the basic ecosystems services (such as drinking water) without which local populations cannot survive
- Organization, particularly the social networks, organization, and institutional systems that enable people to organize responses as constraints or opportunities emerge
- Adapted infrastructure, particularly the design of physical structures (for water, transport, communication, etc.) in ways that can maintain their basic structure and function regardless of changes in climatic systems
- Skills and knowledge, in particular the ability to learn and the basic educational skills required to shift livelihood strategies as required
- Asset convertibility, the development of assets or markets that enable populations to transform the nature of assets and their uses as conditions evolve
- Hazard-specific risk reduction, the development of early warning, spatial planning, implementation of building codes, establishment of community DRR [disaster risk reduction] organization, and other systems to reduce exposure and vulnerability to know climate-related hazards

Source: From Marcus Moench, "Adapting to Climate Change and the Risks Associated with Other Natural Hazards: Methods for Moving from Concepts to Action," in *Adaptation to Climate Change*, ed. E. Lisa F. Schipper and Ian Burton (London: Earthscan, 2009), 273.

1. Identify local climate change impacts
2. Assess community vulnerability
3. Assess local adaptive capacity (local resources)
4. Choose and prioritize adaptive strategies
5. Program and fund implementation

Each of these steps is discussed in the sections that follow.

Identify Local Climate Change Impacts

The first step in adaptation planning is identifying the climate change impacts that may occur locally and developing a projected impact pro-

Box 6.3
Informational Needs for Primary Climate Impacts

For each impact, critical information should be determined: how soon significant or irreversible change may begin, the degree of change from current conditions, and the speed with which the change may occur.

Temperature:
- Duration and frequency of high-heat days and/or heat waves
- Duration and frequency of cold events
- Timing and duration of seasons

Precipitation
- Duration and frequency of drought
- Alteration in annual precipitation total and form (e.g., snow vs. rain)
- Intense precipitation events (e.g., days with total rainfall over a threshold)

Sea Level and Extreme Weather:
- Flood level and frequency
- Level and frequency of extreme high tide
- Frequency and magnitude of extreme weather events

file that answers questions as to how soon significant or irreversible change may begin, the degree of change from current conditions, and the speed with which the change may occur. The increase in average global temperature has altered global climate patterns. The changes in climate can potentially cause a set of projected outcomes, including changes in precipitation pattern, local temperatures (including extreme heat), and sea level rise. These changes directly impact the severity, frequency, and likelihood of other natural hazards or outcomes such as fire, flooding, drought, and species migration. Climate change will not impact all communities in the same manner or to the same degree. The first step for a community is to identify the consequences of climate change most likely to occur locally (box 6.3). Unfortunately, this can be easier said than done.

Characteristics of projected impacts are critical for developing and prioritizing adaptation strategies. Forecast changes that result in large differences from current conditions are more likely to be those climate impacts for which a community is least prepared. For example, in California, climate change is anticipated to increase the probability of large wildfires in some parts of the state.[7] A small increase in wildfire in southern California does present a risk to the region; however, this area already faces frequent wildfire danger and has, in place, emergency response, building standards, and public outreach.[8] While fire occurs infrequently in regions of northern California, the increase in wildfire due to climate change that is projected marks a considerable change from the current condition.[9] Unlike southern regions of California that can adjust or bolster existing policies, the northern regions will need to make large changes to their planning approach, including emergency response, land use planning, and building requirements.

Projected climate change impacts are most often reported at global, national, state, or regional scales and not at local scales. The challenge for local jurisdictions is to identify the climate impacts and magnitude of the changes to expect. International and national entities may have projections specific to the region in which a community is located, but it is often best to first seek climate science reports supplied by state and regional entities such as emergency management and natural resources agencies. For example, the State of California has developed an interactive website to explore state-generated climate change projections to support local jurisdictions seeking to develop adaptation strategies.[10] As of early 2011, twelve states had adaptation plans completed or in progress, and eight others were considering preparing them (fig. 6.3). In the absence of or to complement these reports, national and international data sources can be examined (see the resources section at the end of the chapter). From these reports, a community should seek to gain an understanding of the impacts that will be experienced locally. These data are uncertain, particularly at smaller spatial scales. To account for this uncertainty, a community should identify a range of possible outcomes defined for various future dates such as 2020, 2050, and 2100. In each case, the data and identified thresholds must be tailored for local contexts.

Together, the primary climate change impacts (temperature, precipitation, and sea level rise) can result in a suite of other landscape-scale biophysical outcomes, including wildfire, species migration, and inland flooding. Accurate estimates of these factors can be supported through collaboration with scientific organizations. One of the challenges of

Figure 6.3   Map showing states with climate adaptation plans completed, in progress, or planned (as of February 2011). States with completed plans or plans in progress are shown in gray. Those with the specified intention to create plans are shown in hatch.

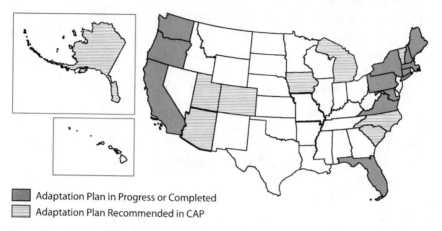

■ Adaptation Plan in Progress or Completed
▥ Adaptation Plan Recommended in CAP

Source: List of states from http://www.pewclimate.org/what_s_being_done/in_the_states/state _action_maps.cfm.

planning for climate impacts is the fact that the climate change science is continually evolving (see appendix A). As a result, adaptation policy must be similarly dynamic and able to be updated as new findings become available. The need to adapt to updated science first requires that communities have identified critical data sources against which to check the projected local changes on which their policies have been based.

Assess Community Vulnerability

The impacts of climate change can set off a cascade of consequences in a community, including alteration of biophysical conditions, safety, human health, and social, economic, and cultural stability. A vulnerability assessment is a systematic way to identify these consequences and evaluate the level of risk presented to a community. For each potential impact such as flooding, fire, or changes in water supply, the potential risks to the community must be assessed. Natural hazards planning provides one potential framework for identifying community resources that may be endangered.[11,12] These impacts can generally be broken down into three broad sectors: *built environment*, including infrastructure, physi - cal damage to private structures (homes and businesses), schools, and

critical facilities; *economic and social,* including lost jobs, business inter-
ruptions, reconstruction costs, shelter requirements, and populations
exposed to impacts; and *ecological* such as loss of biodiversity, insect outbreaks, species migration including agricultural losses, fisheries viability,
and water quality (box 6.4). These sectors are not independent and
often interact. For example, sea level rise may endanger critical transportation infrastructure in coastal areas, which has the potential to
disrupt evacuation routes, posing a threat to public safety.

The following list of sectors and box 6.4 represent a common
suite of considerations. The chance of impact occurring in any one sector will vary by community. As a result, the first step of vulnerability assessment for a community is identifying the particular sectors likely to
be impacted by climate change.

Built Environment: Infrastructure
Transportation infrastructure serves a community by providing connections between homes, work places, and basic services; allowing delivery
of goods and services; and facilitating evacuation from hazards. Disruption of transportation networks, which includes roadways, marine ports,
airports, and train lines, must be carefully evaluated. Damage or loss of
these networks can drastically impact the local economy, isolate populations, and endanger community members. A community should evaluate threats to critical links and nodes in its transportation infrastructure.
Some of these threats may fall into the category of adaptation strategies
that require action despite uncertainty and low probability simply because the risk is too high. A means of assessing such impacts is described
in the section on developing strategies.

Transportation networks can be directly impacted by climate-
related hazard events such as flooding, as well as be affected by changing
conditions such as seasonal climate alteration. Temperature extremes or
repeated freeze–thaw cycles can cause roadways to wear more quickly
and in some cases result in roadways and railroad tracks buckling. These
impacts not only reduce safety but also require heightened diligence regarding maintenance. For a community assessing local vulnerability,
identification of climate impacts on municipal tasks such as road maintenance is a critical piece of information for devising adaptation strategies. The combination of dry periods and high temperatures also
increases the frequency of wildfires. Fire can disrupt transportation net-

Box 6.4
Sectors to Be Evaluated for Vulnerability

Built Environment

Infrastructure
- Transportation (roadways, airports, marine ports, trains)
- Water and wastewater
- Energy
- Communication

Buildings and Planned Development
- Businesses
- Residences
- Community services (hospitals, schools, fire, police)

Economic and Social Setting
- Public health
- Public safety
- Vulnerable populations
- Economy
- Import/export of goods
- Employment level and security
- Economic flexibility
- Food security

Ecosystem Health
- Forests
- Wetlands
- Marine ecosystems and coastal environments
- Agriculture
- Groundwater
- Surface water (rivers and lakes)

works, and this disruption will have direct consequences on community safety as it may impact evacuation routes. The viability of transportation networks for evacuation must also be evaluated with respect to the characteristics of the populations using them to evacuate. Similar to the combination of heat and drought producing fire, intense rainfall can cause floods and landslides that also can disrupt transportation.

Components of a community's infrastructure located in coastal areas, including transportation, water infrastructure, and coastal energy plants, are potentially vulnerable to the impacts associated with sea level rise. Because it is more efficient for wastewater to be conveyed via gravity flow, many water reclamation facilities are located in low-lying coastal areas. Sea level rise may necessitate protective measures to be implemented or, in some cases, relocation of the facility. Either of these measures requires substantial investment of municipal funds. Identification of critical infrastructure with careful assessment of climate science and local conditions allows communities to take early steps to prepare for these impacts. Such preparation can lighten the eventual financial burden and lower community risk. Similar to water reclamation facilities, energy plants are frequently located in coastal areas because they commonly use seawater for cooling. The widespread impact of operations disruption at an energy plant may be a high enough risk to demand action regardless of probability and/or uncertainty. In addition to the possible threats to the structural integrity of energy plants, climate change can impact the efficiency of energy transmission and energy demand. Hot weather or prolonged periods of extreme heat not only result in increased energy demand due to air conditioning use, but also lower transmission efficiency.[13] Increased demand and reduced efficiency may have a range of other outcomes such as brownouts that will impact nearly all aspects of a community. The trickle down consequences of climate impacts on energy supply and demand must be carefully evaluated with particular attention paid to those populations, services, and business that may be most affected.

Built Environment: Buildings and Planned Development
In addition to infrastructure, homes, businesses, and structures that hold critical community services can also be vulnerable to climate impacts. Rapid snowmelt from occurrences such as rain on snow or intense rainstorms can cause the water levels in rivers and streams to rise quickly and increase the probability of flooding. In this case, a community must evaluate the presence of vulnerable structures in potentially flooded areas, the human safety risks associated with flooding, and the vulnerability of structures such as dams or levees. Heavy rain can flood road-

ways, private property, and buildings, particularly basements. The structures that may be subject to physical damage due to flood waters or inundation should be identified and mapped. The structures that may be at risk can hold important economic (including employment), cultural, or safety (e.g., hospitals) roles in the community, which results in loss well beyond that of a structure (as discussed in the next section).

A vulnerability assessment must not only assess existing buildings, but future planned development. Comprehensive plans and zoning codes designate undeveloped areas for future growth. If future development is planned in areas that may be vulnerable to fire, flood, landslide, or other climate impacts, these are local vulnerabilities until the plans are adjusted. Clearly defining the risks posed to planned future development allows communities the opportunity to preemptively develop measures to address climate change through adjusted building standards or by changing the planned location of future growth.

Economic and Social Setting

Economic and social setting refers to the manner in which climate impacts affect public health, safety, and economic stability. Vulnerability in this sector can be produced directly or as a consequence of impacts in other sectors such as infrastructure. When assessing the vulnerability of the built environment, particular interest should be paid to structures that serve critical community functions such as hospitals, schools, cultural centers, emergency services, and critical business hubs.

For many impacts, it is not just the location of structures that may place populations at risk, but the quality of the structures. Coastal counties only constitute 17% of U.S. land, yet over half of the population lives in these areas.[14] As a result, sea level rise and the associated impacts have the potential to threaten the safety of large numbers of people. The vulnerability of these residents should be assessed based on proximity to projected flood impacts and their ability to evacuate. Buildings can also serve as shelter from climate impacts such as fire and heat. However, some portions of a community's housing stock may increase occupant exposure to climate impacts. Housing units can be poorly ventilated or have high sun exposure without having access to cooling such as air conditioning. These living quarters have the potential to amplify the health consequences of heat exposure. Similarly, some structures are more vulnerable to fire than others. If a community is projected to experience increased fire frequency, an evaluation of roofing materials, landscaping, and other fire-resistant building practices should be conducted. An evaluation of a community's building stock should be conducted to identify structures that place inhabitants' safety or health at risk.

Due to work, housing, and other factors some members of a community will be more vulnerable than others. These populations should be identified. Vulnerable transportation infrastructure may limit evacuation options for particular subpopulations in the face of coastal flooding, storms, or fire. The human safety risk of infrastructure vulnerability should be clearly defined. Such an assessment must account for the typical mode of transportation used by vulnerable populations. For example, open roadways facilitate evacuation if the majority of households own a car.

Public health can also be vulnerable to climate change. A change in climate can alter water quality, cause septic system overflows, and foster increased pest populations. In each case, it is an additional source for the transmission of infection, illness, or disease. The potential physical locations and causes of these vectors should be identified. In addition, the populations most likely to be vulnerable to these threats should be identified and their potential exposure evaluated.

Extended periods of high temperatures such as heat waves can result in a variety of health impacts, including severe sunburn, physical weakness and decreased energy, heat stroke, and even death. Heat is particularly dangerous for vulnerable populations such as the young, old, or immunocompromised. Globally, heat waves have resulted in a large number of deaths in the last several years, such as those in Moscow,[15] Paris,[16] and Los Angeles.[17] These instances highlight the need to identify vulnerable populations when evaluating the risk presented by heat. In addition to those populations that may have compromised health, those community members who do not have a means to moderate temperature at home through air conditioning or those who work outside, such as construction or agriculture workers, are also disproportionately vulnerable to heat. High temperature also increases the rate at which ground-level ozone is formed, a priority air pollutant that requires nitrogen, volatile organic compounds, and sunlight to be produced. Ozone has been associated with a wide range of respiratory ailments[18] and should be of particular concern for urban areas due to the combination of increased ozone precursors, the urban heat island effect, and increased heat due to climate change.

The variety of impacts associated with climate change from structural impacts to roads and homes to direct threats on physical safety have the potential to be detrimental to community cohesion. Community cohesion is a critical component to effective response to hazards. The danger of societal fracturing is particularly true when an impact is experienced by only a small portion of a community. In this case, it can create isolation or resentment. It is important for a jurisdiction to identify the potential for these divisions. This can start with evaluation of the access that community populations have to emergency notifications and direction. The evaluation should account for language, access to various media types (e.g., radio, Internet, television), and identify cultural meeting places for all populations within a community.

Effects on local economic stability can result from disruption of transportation networks, changes in resource availability, and changes in employment base. A community should first evaluate the local business sector to evaluate the impact of climate change on the viability of these businesses from the perspective of both local employment and the provision of local goods and services. A business community dominated by a particular sector such as agriculture or trade may be particularly vulnerable. Any number of climate change impacts can have detrimental impacts on a local economy such as disruption of transportation networks that allow for the flow of goods necessary for business. Climate impacts can also impact the financial viability of business through changes such as reduced availability of water, increased temperatures that alter crop productivity, or coastal erosion that reduces the recreational value of a tourist beach.

Ecosystem Health

Projected climate impacts have the potential to drastically change the functioning of ecosystems. With altered precipitation patterns and increased temperature, the annual hydrograph and water quality of rivers and streams will be impacted by climate change. Alteration in flow levels can alter both in-stream habitats and riparian areas. Lower flows and longer low-flow periods can influence fish passage and the viability of water-dependent flora and fauna. Changes in water temperature can also influence habitat conditions. Increased temperature may result in cold-water fish losing habitat and warm-water species expanding their range. Temperature can impact disease vectors and provide improved habitat for insects. Upstream flooding and higher tides may result in more frequent or permanent inundation of coastal habitats. This can alter coastal estuaries and wetlands, dune habitat, and nearshore stream and riparian habitats. Species such as migratory birds, shellfish, anadromous fish, and native plants can all be impacted by shifts in habitat. These changes can also impact commercial fishing and shellfish operations. Another economic role of coastal ecosystems is tourism. The erosion and/or loss of coastal habitats can damage beach recreation and tourism. As sea level slowly rises the first impact may be increased maintenance costs of these areas, particularly those that serve tourism or coastal industries and harbors.

Agricultural growers must match their management practices, including crop choice, tillage practices, planting, harvest, and grazing density, to local climatic conditions. Changes in these conditions such as amount and timing of precipitation, temperature, and the timing of seasons can have detrimental effects on crops and livestock operations. Changes can cause crop damage or failure, new weeds, expanded ranges for existing weeds, new diseases and pests, and damage from extreme rain events or flooding of agricultural areas. These impacts all have the potential to result in reduced yield, which has consequences for the grower, agricultural employees such as field-workers, related industries, and the community at large. Climate change will also impact livestock operations from the stress on animals that results from extended periods of high temperatures or limited water supply. These stressors can result in increased vulnerability to disease and can limit the productivity of livestock operations.

In addition to affecting overall forest health, climate change can leave forests more vulnerable to threats such as insects and fire. The overall impact of climate change on forest productivity varies by location. In some fortunate places, forest productivity may increase. In many others, forest productivity will decline. A decline in forest health can be tied to changes in precipitation patterns and temperature that may alter the timing of rain events, the duration of drought events, and the timing of spring snowmelt. These changes can result in increased tree mortality, species migration, invasive species, pest outbreaks, and changes in interactions between competitive species. Forests stressed by high temperature and limited water are more vulnerable to fire. Large wildfires are projected to increase, and areas with a historically low probability of experiencing fire may be more vulnerable in the future. Damage to forest ecosystems can have economic as well as human health and safety risks. Communities must pay particular attention to the wildland–urban interface.

Assess Local Adaptive Capacity

The complement to the inventory of local risk from climate change impacts is an inventory of the resources and barriers for reducing vulnerability and adapting to the projected changes. These resources include existing policy, local expertise, the capacity for technological innovation, flexibility in the economic base, and high levels of community cohesion. It is important for communities to begin their adaptation strategy development with the resources available locally because actions that draw directly on these resources are most likely to be implemented quickly and supported in the long term because they are less likely to rely on outside help.

The vast majority of communities have existing plans or strategies in place to address some of the potential impacts of climate change. The methods used to develop these existing plans are likely appropriate for some of the impacts that have not yet been addressed, which will build on local expertise. As a result, the first step in evaluating local capability for dealing with local impact is existing plans. Local hazard mitigation plans and comprehensive plan safety elements are the best starting points. In some cases, communities may also have policies that address hazards such as forest management plans or building codes for homes at the wildland–urban interface. Existing plans should be evaluated for the extent to which they address the impacts identified as part of the assessment described earlier. This evaluation acts as a gap analysis that identifies areas of need in existing adaptation strategies. The other benefit of building on existing policy is that strategies that have proven locally effective can be identified. The final piece of assessment in the evaluation of existing policy is whether simply strengthening policy is enough in the face of the projected climate impacts. In some cases, climate impacts will be enough of a deviation from current conditions as to require new strategies.

Addressing climate change impacts requires updating existing policy and developing new strategies and programs. Devising and implementing these strategies requires knowledge, funding, collective

Box 6.5
Managed Coastal Retreat at Surfer's Point, Ventura, California

For coastal communities, the beaches play a critical role defining community identity and serving as a recreational and tourist destination. Coastal erosion threatens to undermine this resource. The City of Ventura, California, has taken a proactive, "managed retreat" strategy that is being hailed as a model for other sites at risk due to climate change.

At this site, a coastal bike path ran along the ocean side of a fairgrounds parking lot near the surf break for which the area is named. Winter storms had scoured enough sand from the site that chunks of asphalt had been washed away. The initial solution suggested for the site was a buried sea wall. Environmentalists and the surfing community argued that this would alter the point break and would sacrifice a beach to protect a parking lot. The debate was settled by the fairgrounds agreeing to give up some of its property to allow for the bike path to be moved 65 feet inland. This solution allowed for the seasonal shift in sand and maintains the natural dynamics of the ecosystem. The City estimates that it will give the site at least 50 years before it faces erosion challenges again.

Source: T. Barboza, "In Ventura, a Retreat in the Face of a Rising Sea," *Los Angeles Times,* January 16, 2011, http://www.latimes.com/news/local/la-me-surfers-point-20110116,0,5658115,full.story.

community action, and, in many cases, innovation. It is important for communities to clearly understand the local resources for addressing the needs identified through impact assessment and existing policy evaluation. In order to prioritize strategies, the ease or difficulty of taking adaptive action should be clearly understood. Elements contributing to local capability can include scientific expertise that can aid in the local interpretation of climate science over time and community organizations that can aid in outreach. Infrastructure concerns can benefit from the capability of local utility providers to make facility adjustments given changing climatic conditions. Box 6.5 describes an effort in Ventura, California, to work with local landowners to relocate a coastal bike path that allows for preservation of the beach ecosystem and the surfing break that serves as a local recreational attraction.

Choose and Prioritize Adaptation Strategies

An inventory of the potential local climate change impacts, the local resources available to address the impacts, and the current deficiencies provides the context necessary to devise adaptation strategies. This process can be difficult given the intersection of uncertainty, multiple needs (emissions reduction, adaptation, and other community goals), and potentially high costs. Decision frameworks can serve a critical role in balancing needs, handling uncertainty, developing strategies, and prioritizing action. It is useful first to understand the characteristics of adaptation strategies, then to evaluate how the impact, vulnerability, and capacity inventories described earlier can be used to identify those issues that warrant immediate action.

Characteristics of Adaptation Strategies

Policy that anticipates climate change impacts with the intention of reducing future risk is inherently uncertain. In addition, adaptation strategies will vary widely because, in contrast to reduction strategies, which are more likely to provide equal benefits to stakeholders, the benefits of adaptation tend to be more spatially explicit. For example, coastal residents will disproportionately benefit from policy focused on adapting to sea level rise. The following list covers some key characteristics of effective adaptive policy:[19]

- *Flexible*: Because climate science is evolving and uncertain, adaptive policy should be robust—that is, applicable under a wide range of conditions. This also implies policy enacted with the assumption that implementation and/or direction will be adjusted over time. Taken to an extreme, the idea of flexibility can be seen as the reversibility of a policy if conditions change or implementation produces unexpected outcomes.
- *Cost-effective*: The benefits of adaptive strategies may not be realized for many years, if not decades. In an economic modeling sense, the further out the benefit, the lower the current value. One way of avoiding this potential conflict between current cost and future benefit is to seek adaptive strategies that have both long-term and short-term benefits or serve as both reduction and adaptation strategies.
- *Specific*: Uncertainty is most easily evaluated in the context of a narrow issue in need of resolution. Climate impacts that require adaptive policy have a projected speed of onset, rate of change, and scale. Policy will be more effective if tailored to address these impact characteristics.
- *Integrative*: Climate impacts have the potential to initiate outcomes in a community across many sectors. Strategy development will be most ef-

fective if the interrelated nature of climate impacts is recognized. Climate change acts directly on things like temperature and precipitation, but adaptive policy may focus on secondary impacts such as the change in temperature and precipitation on crop yield. This policy may be a change in agricultural regulation that will facilitate adaptive change.

Identifying Policy Needs

Once data have been collected (on impacts and local capacity), they need to be organized in a manner that allows for evaluation of priority. One of the first steps is to create a table that begins with data from the impacts assessment. This entails listing the changes that will result from climate change such as alteration of temperature regime, listing the likely local climate consequences, followed by the community resources that may be at risk (table 6.1). These tables can be simple or exhaustive, with consequences broken out by sector or subsector (e.g., fields for impact to transportation infrastructure, public health, etc.).

Compiling all available data provides a clearer picture of the challenges presented by climate change on a local scale. It does not provide a systematic means of identifying areas most in need of attention. The range of impacts and other factors to consider, such as social impact, funding, and co-benefits associated with climate change, necessitates use of a decision framework that will facilitate systematic assessment. The use of a framework also improves the transparency of the process. It can be difficult to communicate the considerations in adaptation policy development. Explicitly defining the decision process and articulating the criteria provides insight and improves community understanding. The first step in this decision process is identifying those impacts that warrant immediate action versus those that can wait. A risk matrix that balances the probability of an impact occurring against the magnitude of the impact is a good way to differentiate between the potential impacts (fig. 6.4). The likelihood of an event should be derived from the current, best available climate science. The assessment of magnitude of impact must reflect local conditions. This is where a committee of local stakeholders should develop criteria that allow for comprehensive evaluation of vulnerability. The City of Keene, New Hampshire, defined criteria in three areas: impacts, influence, and investment (box 6.6).

In many cases, an impact will be addressed through a suite of complementary strategies. Once an impact has been determined to be a

Table 6.1 Example of a local assessment of climate impacts, vulnerability, and capacity

	Local climate impacts	Local vulnerability	Existing local resources
Temperature changes	• Increase in number of days >90°F • Increase in frequency of heat waves • Warmer/shorter winter	• Decrease in air quality due to ground-level ozone levels • Increase in heat-related health issues and mortality • Amplification of urban heat island • Crop damage • Increased water demand • Increased energy demand • Increased wear on roads and rail networks • Increased pest populations (e.g., mosquitoes)	• Air quality testing and notification system • Public education program on heat danger • Insect management program (including spraying) • Water and energy efficiency incentive programs
Precipitation changes	• Increased probability of drought • Increased frequency of high rainfall events • Decreased total rainfall	• Decreased crop productivity • Increased water use • Flood damage to roadways, structures, and agriculture • Combined-sewer overflow • Septic system overflow • Flooding of streets and basements • Decreased water quality • Increased wildfire • Damage to wetland and river ecosystems • Decreased game fish population	• Rain and hydrological monitoring system • Water quality monitoring • Fire safety education programs • Fire safety, building standards • Drought-tolerant planting requirements
Sea level rise	• Increased extreme high tide level • Increased flood frequency 2-, 10-, and 100-year floods (e.g., estuaries)	• Seawater intrusion into groundwater resources • Coastal inundation • Erosion of coastlines • Loss of coastal ecosystems in coastal zone • Damage to recreational settings • Flooding or structural damage to coastal buildings and infrastructure • Danger to coastal utilities (electricity, wastewater)	• Groundwater monitoring and mapping • Natural shoreline protection policy • Development restrictions

Figure 6.4   Scenario A illustrates a risk matrix that can be used to decide which climate impacts warrant development of strategies. The second use of the matrix, scenario B, demonstrates how availability of funding and timeliness can be balanced to define near-, mid-, and long-term strategies. Similar matrices can be used to balance a wide range of climate impacts, community needs, and feasibility constraints.

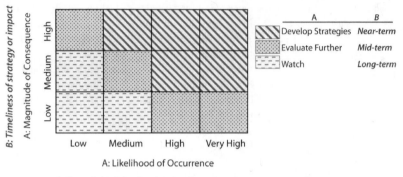

high priority, the host of potential strategies identified to address the impact should be evaluated and prioritized in the manner similar to that suggested in chapter 5 for emissions-reductions strategies. Some strategies may be delayed due to factors such as reliance on technological advancement. A decision matrix can be used to aid in balancing implementation considerations (see fig. 6.4). Similar to the assessment of risk, it is up to a community to define costs and benefits and evaluate the timing given local resources. Costs may be financial, but they can also be in social, ecosystem, or other sectors. Benefits could refer to strategies that meet community goals in addition to adaptation.

Implementing Adaptation Strategies

Similar to reduction strategies, part of developing adaptation strategies is identifying a responsible entity, phasing program, funding source, and monitoring program (discussed in detail in chap. 7). In climate adaptation two additional factors are critical components of long-term policy implementation. A climate adaptation strategy must be responsive to changes in climate science and be easily adjusted based on the nature of experienced climate impacts or hazards and strategy implementation effectiveness.

Box 6.6
Adaptation Planning in Keene, New Hampshire

The City of Keene divided its climate planning into two free-standing plans, one for reduction strategies and one for adaptation strategies. The plan, *Adapting to Climate Change: Planning a Climate Resilient Community*[a], organizes the development of adaptation strategies into six sections: why plan to adapt to climate change; climate change impacts; assessing community climate vulnerability; climate adaptation priority goals and targets; moving forward; and lessons learned. The chapters focus on three community systems: the built, natural, and social networks.

The plan was developed by a committee consisting of city officials (mayor and city council members), local and regional planning staff, city department heads, representatives from the college community, and local stakeholders from the public health field. Representatives from national and regional science and policy entities also supported the process. Vulnerability to impacts was organized into sectors, evaluated for a range of consequences, and each outcome was then rated. The rating system had three levels: *no impact*, opportunity, or contribution to the community; *minor impact*, opportunity, or contribution to the community; and *great impact*, opportunity, or contribution to the community.

Vulnerable Sectors and Subsectors

Built Environment
- Buildings and development
- Transportation infrastructure
- Stormwater infrastructure
- Energy systems

Natural Environment
- Wetlands
- Groundwater
- Agriculture

Social Environment
- Economy
- Public Health
- Emergency Services

Rating Criteria

Impacts:
- Local business
- Environment
- Community

Influence:
- Visibility
- Supporting existing initiatives

Investment:
- Available funding
- Easy to implement
- Time sensitive
- Cost-effectiveness

Highlights of Plan Development

The assessment of potential climate impacts conducted for the Keene plan relied on regional and state-level findings reported by the Union of Concerned Scientists. The impacts were categorized by season with information on timing and length, temperature (extreme heat, ice melt), and precipitation (snowfall, drought, flooding). Local vulnerability to these climate changes was identified based on the sectors and subsectors listed above. The analysis explicitly recognized the overlap between the various sectors and embraced the interaction. For example, climate threats to communication infrastructure were considered as part of the built environment, but their role in emergency services was recognized as part of the analysis. Each of the vulnerabilities or risks identified was then evaluated based on criteria, and adaptation strategies were defined. To assure implementation, targets were defined for each strategy.

[a] City of Keene, New Hampshire, *Adapting to Climate Change: Planning a Climate Resilient Community* (2007).

The following are some of the factors critical to long-term, effective implementation:[20]

- *Political leadership*: Adaptation strategies are addressing changes that are often going to occur in the future, meaning there are unlikely to be immediately observable benefits of implementation. Strong leadership assures adaptation strategies have an appropriate level of support. Implementation of individual strategies will be conducted by a variety of departments and entities, from fire departments to utility directors. A single agency or interagency group should be charged with coordinating implementation through time and across spatial scales. This leadership is seen as a critical step in facilitating integration of policy with the intent of mainstreaming climate adaptation into everyday decisions.

- *Funding:* Adaptation strategies are not identified as frequently as reduction strategies and are rarely funded independently. Increased emphasis on anticipatory adaptation will require funding levels similar to emissions reduction. The funding need not be entirely separated from reduction strategies, but some level of distinction should be made to assure that adaptation strategies are developed and implemented. One possible mechanism is an incremental increase in existing reduction funds and application of these funds to adaptation based on the premise that effective adaptation reduces vulnerability.

- *Development and diffusion of science and technology:* Adaptation relies on predicted climate impacts. The uncertainty associated with these impacts increases at smaller scales. As a result, there must be a tight feedback loop between science and technology and policy development. Not only does there need to be consistent communication, the diffusion of information will require mechanisms put in place to assure consistency throughout the community, including public education. There is also an opportunity for communities to forge relationships with local science entities. This may allow for collaboration to generate science that specifically fills a local need.

- *Feedback loops and adaptive policy:* This can be seen as another aspect of climate science development and diffusion. Climate adaptation policy must have indicators that allow for evaluation of policy success. These data in combination with advances in climate science and development of new technology must inform revision of adaptation policy. Periodic review and a mechanism for policy revision should be established as part of the strategy development process.

Chapter Resources

Climate Change Impact Assessments

Thomas R. Karl, Jerry M. Melillo, and Thomas C. Peterson, eds., *Global Climate Change Impacts in the United States* (New York: Cambridge University Press, 2009). http://www.globalchange.gov/publications/reports /scientific-assessments/us-impacts. "The report summarizes the science and the impacts of climate change on the United States, now and in the future. It focuses on climate change impacts in different regions of the U.S. and on various aspects of society and the economy such as energy, water, agriculture, and health. It's also a report written in plain language, with the goal of better informing public and private decision making at all levels."

Pew Center on Global Climate Change, *Climate Change 101: Understanding and Responding to Global Climate Change* (Arlington, VA: Author, January 2011), 94 pp. http://www.pewclimate.org/climate-change-101. This report series includes summaries of climate science, summaries of the current state of practice in the United States, and basic guidance on climate adaptation policy development.

Guidance Documents

A. K. Anover, L. Whitely Binder, J. Lopez, E. Willmott, J. Key, D. Howell, J. Simmonds, *Preparing for Climate Change: A Guidebook for Local, Regional, and State Governments* (Oakland, CA: ICLEI–Local Governments for Sustainability, 2007), 186 pp. www.icleiusa.org/action-center /planning/adaptation-guidebook. This guidebook offers a step-by-step process for developing adaptation strategies. Its strength is in the accessible language and thorough treatment of the issue from interpreting climate science to strategy implementation and monitoring.

Federal Emergency Management Agency Hazard Mitigation Planning "How-To" Guides. http://www.fema.gov/plan/mitplanning/resources.shtm. FEMA has prepared nine "How-To" guides that provide step-by-step instructions on how to prepare hazard mitigation plans. They cover getting started, identifying risks and estimating losses, using cost–benefit analysis, implementing plans, and a variety of other topics. Highly recommended for communities who want to assess their risks to hazards and disasters.

Federal Emergency Management Agency, "Local Multi-hazard Mitigation Planning Guidance" (Washington, DC: FEMA, 2008). http:// www.fema.gov/library/viewRecord.do?id=3336. "To help local governments better understand the Local Mitigation Plans requirements under 44 CFR Part 201, FEMA has prepared this document with two major objectives. First, the Guidance is intended to help local jurisdictions develop new mitigation plans or modify existing ones in accordance with the requirements of the regulation. Second, the Guidance is designed to help Federal and State reviewers evaluate mitigation plans from local jurisdictions in a fair and consistent manner."

National Research Council, *Adapting to the Impacts of Climate Change* (Washington, DC: National Academies Press, 2010). http://americasclimate choices.org/paneladaptation.shtml. This is a congressionally requested study from the National Research Council that provides an introduction to the challenge of climate change adaptation. It also provides an argument for why a coordinated response is needed at all levels of government to address the risks presented by climate change.

United Nations Development Programme. "Adaptation Policy Frameworks for Climate Change." http://www.undp.org/climatechange/adapt/apf .html. This United Nations website contains guidance on developing a climate change adaptation plan. It addresses scoping, vulnerability assessment, future climate risks assessment, strategy formulation, and monitoring and evaluation.

Examples

City of Keene, New Hampshire, *Adapting to Climate Change: Planning a Climate Resilient Community* (2007). http://www.ci.keene.nh.us/sites /default/files/Keene%20Report_ICLEI_FINAL_v2.pdf. The City of Keene prepared a stand-alone climate adaptation plan following ICLEI's Climate Resilient Communities process. The plan documents regional climate impacts, assesses local vulnerability, and establishes adaption goals and strategies. It also includes a "Lesson Learned" section, which should be helpful to other communities.

New York City Panel on Climate Change, *Climate Risk Information* (February 17, 2009). http://www.nyc.gov/html/om/pdf/2009/NPCC_CRI .pdf. The City of New York has produced this report, which is one of the first steps in acting on the adaptation strategies in the "Climate Change" section of their sustainability plan, *PlaNYC*. It includes climate change scenarios for the city, observed changes and future projections, infrastructure impacts, and indicators and monitoring. It will be followed by two additional reports.

Chapter 7

Implementation

A plan is only as good as its implementation. Too often communities invest considerable effort in preparing a quality planning document only to see little happen because of failure to implement. Climate action plans (CAPs) present an implementation challenge since they are unfamiliar to many, often cut across organizational boundaries, and often lack dedicated sources of revenue. These are all challenges that can be addressed during plan development. This chapter addresses phase III: Implementation and Monitoring of the climate action planning process presented in chapter 2. CAPs should include a section that addresses how the plan will be implemented. The section should answer the following questions:

- Who will be responsible for oversight and management of the CAP implementation?
- Who will actually implement each strategy in the CAP?
- What will be the timeline or phasing (programming) for implementation of the strategies?
- How will implementation of strategies be funded?
- How will implementation be monitored and evaluated?

The chapter concludes with a discussion of whether CAPs should be integrated with local comprehensive land use plans to facilitate implementation.

Implementation Oversight and Management

There are two basic options for organizing the implementation program of a CAP. The first is to create new entities with responsibility to coordinate implementation of the CAP. As interest in climate change and sustainability has grown, many local governments are establishing specific offices or departments to address these issues and creating a sustainability or climate program coordinator position to direct implementation. Some have created "green teams" of cross-agency staff members to coordinate plan implementation. The community should decide where to house these new climate offices. Most communities have chosen to house climate programs in environmental, planning, or public works departments to take advantage of the specialized expertise and similarity of mandates. Some communities have chosen to house them in the mayor's or city manager's office to take advantage of the authority and visibility of those offices.

The second approach to program organization is to rely on the existing organization to implement the CAP. The benefits of this approach are that it requires little or no new staffing, does not disrupt the existing institutional culture, and empowers those closest to actual implementation to take action. The potential problems are that existing mandates and programs may overshadow the new climate strategies, implementation may be uneven across parts of the organization, and accountability may be diffuse.

Implementation Committees or Green Teams

Implementation committees or green teams can be a very effective approach for ensuring implementation of the CAP. If a community decides not to have an implementation committee then the plan should clearly designate a responsible entity for ensuring implementation of each greenhouse gas (GHG) reduction and climate adaptation strategy.

Decisions about how to structure an implementation team should be based on how the plan itself is constructed and the nature of the emissions reduction and adaptation strategies in the plan (box 7.1). Communities with plans prepared by citizen committees or task forces may want to continue or modify those groups to be directly involved in implementation or to serve in an oversight or watchdog capacity over those who will implement the plan. Communities that adopt emissions reduction and adaptation strategies dependent primarily on government action may want to transition their Climate Action Team (CAT) into an implementation team.

Box 7.1
Examples of Implementation Committees

Cincinnati, Ohio's Climate Protection Steering Committee

There are two important aspects of the *Green Cincinnati Plan* (formerly called the *Climate Protection Action Plan*) that affect implementation. First, preparation of the plan was led by the mayor-appointed Climate Protection Steering Committee (CPSC) and assisted by five community-based teams. Due to its success, the CPSC was reestablished and expanded to lead the effort on implementation. Second, mitigation strategies were adopted that required action from businesses and nonprofits in addition to the municipal government. The plan directs the CPSC to designate an entity to spearhead the implementation for each strategy in the plan; this lead entity could be "an individual, a City department, a business, or an organization, depending on the nature of the action being implemented." In Cincinnati the plan is best described as belonging to the community as a whole; thus implementation is thought of as the responsibility of the community rather than just the municipal government, although the municipal government is an important responsible party. The CPSC, made up of twenty-eight community leaders, plays a critical role in ensuring the community responds to its responsibility for implementation. It must also prepare an annual report and regularly update the city council on progress.

Hayward, California's Climate Action Management Team

In contrast to Cincinnati's approach, the City of Hayward's Climate Action Plan contains mostly emissions-reduction strategies to be implemented by city departments and agencies. Hayward's plan establishes a Climate Action Management Team (CAMT) made up of representatives from the city's Finance, Public Works, Development Services, Maintenance Services, and Library and Neighborhood Services Departments as well as the city manager's office. The plan directs the City to consider representatives from key city committees and commissions. The CAMT membership reflects the nature of the emissions-reduction strategies. Because the members are city employees, and perhaps a few appointed committee members, they answer directly to the city manager and city council. They will also have direct influence over funding and staffing decisions that affect plan implementation.

Key Questions for Implementation Committees

- Who will serve on the team and how will they be chosen?
- What will be the role of the team? Will it be mostly oversight, or will the team take direct responsibility for implementation?
- What authority will the team have to ensure implementation? Will the team control funding associated with implementation?

- How will the team be held accountable?
- Will the team have responsibility for outreach and communication?
- Will the team have responsibility for monitoring, evaluation, and progress reporting?
- Are team members subject to state or local sunshine or open government laws for conflict of interest, financial disclosure, ex parte communication, and the like?

Sustainability or Climate Program Coordinators

Some communities hire climate program coordinators or managers, usually as a local government position but sometimes through nonprofit organizations. These climate program coordinators are sometimes brought in to prepare the plans themselves but more often are recommended in plans as a necessary component of a successful implementation strategy. The increasing sophistication and comprehensiveness of CAPs is generating a need for this new professional class. These may be full- or half-time positions and may or may not come with additional support staffing. When hired by the local government they tend to work directly for the top administrative office of the local government—the city administrator/manager or mayor. Examples of these different approaches include the City of Santa Cruz, California, which has a half-time climate change action coordinator housed in the Planning and Community Development Department, and the City of Pittsburgh, which has a full-time sustainability coordinator housed in the mayor's office.

Entities Responsible for Implementing Strategies

The implementation committee or climate program coordinator must identify a specific individual, agency, department, or organization responsible for implementing each strategy. For example, a strategy to replace lightbulbs in traffic lights with high-efficiency LEDs would likely be assigned to a municipal transportation or public works department. Assigning this responsibility may be met with some resistance. Local government agencies and departments may feel this is another burden on their already busy staff. Community partners may not feel they have the knowledge or capacity to implement strategies. This is why chapter 2 suggests that local government and community partners be involved in preparation of the plan from the beginning. They need to buy in to the process and contribute to the development of strategies knowing they may be assigned some responsibility for implementation.

One consideration in implementation is how to hold these entities accountable. This is where having an implementation committee with the right people can be critical to success. If it is staffed by the people who have decision-making authority in their respective agencies and organizations, then it is much easier to ensure implementation. This is usually easier to accomplish with strategies that are to be implemented by local government agencies and departments since they usually have clear lines of authority. This may be more difficult with strategies that are to be implemented by community organizations such as nonprofits, which may have unclear hierarchies or little organizational capacity. In this case, the implementation committee may have to take a stronger role in ensuring accountability.

Programming of Strategies for Implementation

Previous chapters on emissions reduction and climate adaptation strategies discussed strategy evaluation and prioritization. This information should be used by the implementation committee or climate program coordinator to program the priority and timing of strategy implementation (see box 7.2). Typically priority and timing are driven by access to funding or capacity/capability of the implementing organization. For funding, budget cycles, grant funding cycles, and fundraising programs may drive when a strategy can be implemented or how fast it can be fully implemented. For capacity and capability, issues of staffing levels, staff expertise, workload, consistency with current mission, and timing will be important. Other projects may need to be completed first, staff may need additional training to support the program, or some level of reorganization may be needed; each of these can delay action.

Priority and timing may also be driven by external events or circumstances, strategy synergy, and outreach considerations. Sometimes a higher-priority strategy may partially depend on, or be synergistic with, a lower-priority strategy or other community action. In this case it may make sense to deviate from a strict prioritization scheme. In addition, communities should consider which strategies can be piggybacked on existing actions, for example, taking advantage of a road repaving to add bicycle lanes. They should also consider strategies that have high public

Box 7.2
Example of Evaluation Criteria from the City of Oakland, California, Energy and Climate Action Plan

Evaluative criteria	Issues to consider
GHG reduction potential	• Magnitude of GHG reductions • Measurability of reductions
Implementation cost and access to funding	• Cost to City budget • Cost to other stakeholders • Access to funding
Financial rate of return	• Return on investment to City and/or stakeholders implementing the action • Protection from future costs
GHG reduction cost-effectiveness	• Relative cost/benefit assessment in terms of estimated GHG reductions
Economic development potential	• Job creation potential • Business development and retention potential • Workforce development potential • Cost savings to community • Education benefits for community
Creation of significant social equity benefit	• Benefits to disadvantaged residents in the form of jobs, cost savings, and other opportunities • Reduction of pollution in heavily impacted neighborhoods • Equity in protection from impacts of climate change
Feasibility and speed of implementation	• Degree of City control to implement the action • Level of staff effort required • Resources required • Degree of stakeholder support • Amount of time needed to complete implementation • Time period during which implementation can begin
Leveraging partnerships	• Leverage partnerships with community stakeholders • Leverage partnerships on a regional, state, or national level • Facilitate replication in other communities
Longevity of benefits	• Persistence of benefits over time • Opportunity to support future additional benefit

visibility that can serve to educate or motivate the public. Regardless of the rationale, all strategies should be assigned a priority and timing for implementation.

Financing

Finding money for implementing the CAP strategies can be the most challenging aspect of implementation. CAPs must compete against all the other needs in a community, which in difficult economic times can be a problem. This section presents a variety of funding types and examples.

Emissions reduction and climate adaptation strategies should have estimated costs for implementation attached to them. These costs can then be assembled into a budget based on the prioritization and timeline that have been established. In addition to the costs for each strategy, there should be a budget for overall program administration that includes staffing, education and outreach, plan monitoring and updating, and so forth.

Sources of Funding

The following are potential sources of funding for the CAP (box 7.3):

- General funds
- Bonds
- Taxes and fees (impact fees)
- Government grants
- Carbon offset programs
- Self-funding and revolving fund programs
- Volunteer and pro bono resources
- Private grants
- Private investment

General Funds

A community may choose to allocate a portion of the local government's general funds to implementation of the CAP. The general fund is the local government's primary operating account with revenues that are usually generated from property taxes, local sales taxes, and other local taxes and fees. Assigning general funds can be challenging because

Box 7.3
Example of Funding Source Identification: Homer, Alaska

The City of Homer included the following in the Implementation section of their Climate Action Plan:[a]

The City of Homer will establish and promote a "Sustainability Fund" which will be used to help cover the costs of implementing the Climate Action Plan.
Possible sources of revenue for the Sustainability Fund include:
- Grant funding from state and federal programs and private foundations.
- A Climate Action Plan tax modeled after Boulder, Colorado's innovative program. The CAP tax in Boulder, approved by voters, involves an agreement with the local investor-owned electric utility to assess a tax for residential, commercial, and industrial customers based on electricity usage. The tax is collected as part of the utility's normal billing process.
- A per-gallon tax on all fuel transferred within the City of Homer.
- Voluntary "offsets" contributed by individuals and businesses who wish to reduce their carbon footprint by supporting projects aimed at reducing greenhouse gas emissions in the community at large.
- Funds contributed by the City of Homer to offset employee travel (calculated as $X per ton of travel-related CO_2).
- Savings resulting from increased energy efficiency/conservation as CAP measures impacting City operations are implemented.
- Homer Spit parking fees.

[a] City of Homer, Alaska, Climate Action Plan (December 2007), 40.

it usually means shifting them from another community program. Most local governments have tight operating budgets that make assigning them to CAP implementation a difficult proposition.

Bonds

Local governments may issue bonds, essentially borrowing money from the bond holder to finance the CAP. Bonds are usually used for capital projects (e.g., public buildings, roads, sewer and water infrastructure, etc.) or for projects that generate revenue to pay off the bond (pay-parking garages are a typical example). If an emissions reduction strategy in the CAP fits one of these project types then issuing a bond may be a viable approach.

Taxes and Fees
Local governments can initiate new taxes or fees to fund climate programs. Taxes and fees could be broadly applied, such as a sales tax, or they could be tailored to link certain behaviors. For example, in the city of Boulder, Colorado, voters approved Initiative 202 in November 2006, which established the Climate Action Plan Tax (also known as the Carbon Tax). The local utility provider, Xcel Energy, collects a tax on electricity for the city, and the City uses it to fund implementation of the CAP. Not only does the tax raise funds but it also increases the cost of electricity, which should lower its use, thus providing a direct emissions reduction benefit.

Government Grants
Federal and state governments offer a variety of grants to assist local governments in implementing emissions reduction and climate adaptation strategies. The types of grants will need to be monitored year to year since they change frequently. Some grants require matching funds, some are competitive, and some require certain conditions to be met. Communities should investigate these and prepare to satisfy these types of conditions in advance so that they can quickly take advantage of new rounds of funding.

Private Grants
There are numerous for profit and nonprofit organizations that offer grants to support programs that fit into CAPs. Communities should look to their nonprofit partners who likely have expertise and experience in finding and securing grants.

Private Investment
Some climate strategies may attract private investment. Two common examples are car- and bike-sharing services and solar power purchase agreements (SPPAs).

Carbon Offset Programs
Certain activities can be linked to the required or voluntary payment of an additional fee in relation to the amount of GHGs the activity would create. Activities could be driving, flying, disposing of waste, using water, and the like. The idea is to offset the GHGs the activity creates by funding strategies that compensate with an equivalent reduction. This is usually done where the emissions reduction for the original activity itself is

very difficult, or perhaps impossible at the local level. For example, the City of San Francisco created the San Francisco Carbon Fund, which puts a 13% surcharge on all city employee air travel, and set up carbon offset kiosks in San Francisco International Airport (Climate Passport Program) for voluntary payments by the general public. The Fund is in its pilot phase but has already generated funds for a biodiesel project and an urban orchard project. Offset programs are generally seen as a last resort option for emissions reduction but can be a good source of funding for programs.

Self-Funding and Revolving Fund Programs
These are programs established to generate their own revenue and are similar to municipal enterprise funds (box 7.3). Usually the revenue generated is directly from the recipient of the benefits of the program; thus they are usually seen as fair and equitable programs. The City of Berkeley, California, created the Berkeley FIRST program to finance the cost of solar installations through an annual special tax on a home-owner's property tax bill that is repaid over twenty years. The key innovation is that, since installation of solar is an improvement to the property, the loan stays with the house. If the house is sold, the new owner would take over the payments for the improvement since they would reap the benefits of the lower utility costs. The Berkeley FIRST program requires little up-front cost to the property owner and thus creates a strong incentive to install solar. The City of Phoenix, Arizona, created the Phoenix Energy Conservation Savings Reinvestment Fund to provide capital for energy-efficiency projects. As the City invests in energy-efficiency measures, such as installing high-efficiency lighting, it reinvests half of all documented annual energy savings, up to a limit of $750,000, into a revolving fund.

Volunteer and Pro Bono Resources
In many communities, nonprofit and service organizations, businesses, and individuals are willing to donate money and services to important causes. The City of San Luis Obispo, California, has bicycle valet parking year-round at its Thursday night farmer's market. The bike valet is provided at no charge to the community by a local nonprofit and is staffed by volunteers.

Monitoring and Evaluating the CAP

CAP strategies should contain a program for monitoring progress on implementation and achieving GHG emissions reduction targets, a program for reporting and publicizing these achievements, and a process for evaluating and updating the plan.

CAPs should contain a program for monitoring three aspects of implementation:

1. *Basic monitoring*: This includes determining whether the strategy was in fact implemented, met its budget, and was implemented on schedule.
2. *Success of desired direct action*: If the strategy was implemented did it produce the desired effect or outcome? For example, did the number of people expected to install solar panels through a solar incentive program do so, or did the expected increase in transit ridership from an employee rider discount program in fact occur?
3. *Level of GHG emissions reductions*: How much do GHG emissions change and how well does the change meet the adopted GHG emissions reduction target? This third aspect can be considered the bottom line.

Each of these levels of monitoring can be linked to performance indicators that show how well the community is doing in achieving expected levels of performance for each strategy (table 7.1). For example, if a community had identified that it needed to do energy efficiency retrofits on 10% of the houses and businesses in the community to achieve the desired GHG emissions reduction in that sector, then the performance indicator would track and report the percentage of retrofits. These performance indicators can be tracked on a regular basis and reported as a scorecard for implementation of the CAP. This allows the progress of the CAP to be easily communicated so that decision makers can make adjustments to program implementation.

Communities should consider annual or biannual reporting of progress on CAP implementation and GHG emissions reductions. An annual report can be used to inform those who participated in creating and adopting the plan of the progress of their work. In addition, the annual report can serve as an important component of educating and motivating the public about what needs to be done to address the climate change problem. An annual report also helps to ensure that the CAP isn't ignored and holds accountable those responsible for implementation.

Table 7.1 Examples of progress indicators

Climate action strategy	Progress indicator
Develop an energy efficiency financing program (through PACE, Energy Upgrade, or other mechanisms) allowing property owners to invest in energy efficiency upgrades and renewable energy installations for their buildings.	Percentage of households and businesses participating Average electricity savings Average natural gas savings
Revise policies and regulations as needed to eliminate barriers to or unreasonable restrictions on the use of renewable energy.	Megawatts of renewable energy systems installed
Implement a curbside compost pickup in combination with existing green waste pickup.	Tons of food waste diverted and tons of green waste diverted
Amend applicable ordinances and policies to direct most new residential development away from rural areas and to concentrate new residential development in higher-density residential areas located near major transportation corridors and transit routes, where resources and services are available.	Percentage of residents within half a mile of a transit stop
Incorporate Complete Streets policies into the Circulation Element and implement Complete Streets policies on all future roadway projects.	Miles of bike lane and sidewalks installed
Implement tiered water rate structures to incentivize water conservation.	Gallons of water saved Per capita water use reduction
Develop and disseminate appropriate best management practices for the application of pesticides and fertilizers, tillage practices, cover crops, and other techniques to reduce nitrous oxide emissions, maximize carbon sequestration, reduce water use and runoff, and reduce fuel use.	Crop fertilization rates per acre

It would not be reasonable to expect a GHG emissions inventory and CAP update every one to two years; instead communities should consider a five-year update schedule or perhaps tying the update to interim reduction target years. In between the major updates, the community can use the annual report as an opportunity to modify strategies and include interim updates as needed. Press releases can accompany the release of annual reports and updates, which can be made available on websites and public places such as libraries.

Positioning the Climate Action Plan within the Comprehensive Plan

More communities are considering integrating their CAP policies and strategies into their comprehensive land use plan. There are several reasons for this: the comprehensive plan is an existing, recognized legal instrument for implementing and enforcing policies and strategies; comprehensive plan implementation is usually linked to specific departments thus providing ownership and accountability; and there will often be overlap in issue areas between the two plans where synergies can be captured and potential conflicts resolved. In the long term, complete integration of climate action strategies in community planning documents may become standard. At present, integration of the CAP into the comprehensive plan is not required, though it is an increasing trend, often to ensure long-term implementation. Integration with the comprehensive plans requires consideration of the ways in which a CAP differs to assure that the two policy documents are consistent and complementary.

Plan Content and Structure

The content and structure of comprehensive plans will vary based on state law or local preferences. An accepted hierarchy has evolved to include a vision, planning principles, goals, objectives, policies, actions or implementation measures, and indicators or performance measures. Comprehensive plans include multiple components or elements to address key issues, such as land use, transportation/mobility, public utilities and infrastructure, safety/hazards, noise, housing, agriculture, open space, energy, air quality, water resources, biological resources, historic preservation, cultural resources, public health, parks and recreation, and economic development. Comprehensive plans often balance land use, social, environmental, and economic objectives.

Plan Integration

Communities should identify their key goals related to climate action and then develop, update, or amend their comprehensive plans to reflect those goals accordingly. Integration of climate action into a local comprehensive plan will be most straightforward when the plan is being updated or developed concurrently with or following the development of a stand-alone CAP. Incorporation of climate action goals, policies, and actions into a local community plan should occur across the various elements of a comprehensive plan rather than into a single element.

- *Land use element*: With regard to land use, the most important actions to reduce GHG emissions discourage auto-dependent, low-density development, and promote complete communities that provide mixed land uses, higher densities in core areas and transit nodes, affordable housing, compact form, smart growth, and bicycle- and pedestrian-friendly infrastructure.

- *Transportation, circulation, or mobility element*: Transportation-related policies should discourage vehicle miles traveled, specifically travel by single-occupant motor vehicles, and encourage the use of mass transit, bicycling, walking, and telecommuting. Local governments should accommodate all modes of transportation and make them attractive and convenient. Local streets should accommodate users of all transportation modes throughout the community; this is often referred to as complete streets. Communities should incentivize the use of and fund transit with needed infrastructure, plan for complete networks of bikeways, ensure that streets have safe and inviting sidewalks and street crossings, provide park-and-ride facilities, and plan shared parking where appropriate.

- *Housing element*: The key connection between the housing element and GHG emissions is the community's jobs/housing balance. The housing elements should include policies and actions to ensure a balance between job and housing availability as a means to reduce vehicle miles traveled. In addition, housing elements can include policies on energy conservation or green building and renovation.

- *Conservation, environmental, or natural resources element*: Policies in this element should promote energy, water, and other resource efficiency and conservation.

- *Open space and agriculture element*: Policies in this element should address the protection of agriculture, forests and woodlands, and the expansion of urban parks and street tree programs as these resources serve as carbon sinks or sequestration opportunities.

- *Natural hazards or safety element*: Climate change may increase the frequency and severity of natural disasters such as wildfires, flooding, droughts, and heat emergencies. The natural hazards or safety element can direct adaptation to these changes, for example, by incorporating policies that restrict development in the wildland–urban interface, along shorelines, and on floodplains.

In the long term, it is possible that the goals of reduced GHG emissions and resilience in the face of unavoidable climate impacts will become ubiquitous and a standard component of comprehensive plans. This integration may shift the role and framing of a CAP. A local jurisdiction considering a CAP must evaluate where it lies in this continuum and where it envisions itself in the long term.

Chapter Resources

Books

Eugene Bardach, *A Practical Guide for Policy Analysis: The Eightfold Path to More Effective Problem Solving* (Washington, DC: CQ Press, 2004). The book provides a basic, straightforward approach to evaluating proposed public policies. It can be used to inform developing and applying evaluation criteria to emissions reduction and climate adaptation policies.

Examples

City of Seattle, Washington, *Seattle Climate Protection Initiative Progress Report.* http://www.cityofseattle.net/climate/. The City of Seattle prepares an annual progress report that includes the status of specific strategies and a progress report on GHG emissions reductions. The report is full color and designed to be accessible to the general public.

City of Portland, Oregon, *Local Government Publications on Climate Change.* http://www.portlandonline.com/bps/index.cfm?c=41917. The City of Portland has not only completed three climate plans but also interim progress reports that summarize strategy effectiveness based on monitoring. This site provides PDF links not only to the plans but also to each of the monitoring reports.

Chapter 8

———————— ✦ ————————

Communities Leading the Way

This chapter presents six cases of communities that have prepared climate action plans (CAPs) and are now in the process of implementing those plans. The cases are chosen for their diversity of experiences and lessons learned. They illustrate many of the principles outlined in this book and demonstrate that climate action planning is possible in all types of communities. The City of Portland and Multnomah County, Oregon, have been in the business of developing and implementing CAPs since the early 1990s and show how to construct a successful program over the long term. The City of Evanston, Illinois, shows the benefits of building "social capital" in the community that creates a grassroots capability for doing community-based climate action planning. The City of Pittsburgh, Pennsylvania, demonstrates the power of local partnerships among public, private, and nonprofit entities to develop and implement plans. The City of San Carlos, California, shows how a city-led planning process integrated into an update of the city general plan can ensure that a CAP will have the authority and backing to be successfully implemented. Miami-Dade County shows that counties can do climate action planning and that it can be integrated with a larger effort of achieving sustainability. And finally, the City of Homer, Alaska, demonstrates that big ideas can come from small places and that implementation is where the real work takes place. Communities beginning to work on their own climate action plan can look to these communities for insights on how to best prepare a CAP.

City of Portland and Multnomah County, Oregon: Two Decades of Leadership[1]

The City of Portland, Oregon, has been a leader in climate change policy development for almost two decades. The City adopted the first carbon-reduction plan in the United States in 1993. Since that time, Portland has forged a collaborative relationship with Multnomah County and has twice revised this original plan (2001 and 2009). These revised plans had the benefit of learning from the successes and challenges of earlier efforts. The sustained and gradual effort by Portland to reduce greenhouse gas (GHG) emissions provides critical lessons for cities that are much earlier in the climate planning and implementation process.

Portland and Multnomah County's plans were developed through collaboration between multiple departments, local organizations, and the public. The success of Portland and Multnomah County's efforts can be tied to the development of strong community partnerships, a commitment that CAPs should be meeting a variety of community goals in addition to GHG reductions, and long-term monitoring that allows strategy effectiveness to be assessed (box 8.1).

The first plan adopted by the City of Portland in 1993 was motivated by the City's participation in the World Conference on the Changing Atmosphere held in Toronto, Canada, in 1988 and the United Nations development of the Kyoto Protocol in 1992. The plan was titled *Carbon Dioxide Reduction Strategy*. The target established in this plan went beyond that established by the Kyoto Protocol; the City aimed to reduce emissions to 20% below 1988 levels by 2010. The reductions required to reach this target were divided among five local action areas: transportation, energy efficiency, renewable resources and cogeneration, recycling, and tree planting. The breadth, level of detail, and ease of implementation of the strategies included in this 1993 plan set the stage for the success of future plans and actions.

Monitoring and evaluation of strategy effectiveness allowed the City to identify areas of success, as well as confounding factors for overall emissions reduction. The 2000 status report, *Carbon Dioxide Reduction Strategy: Success and Setbacks*, individually evaluated the action items in the 1993 plan. Overall, this report showed a reduction in per capita emissions, but increases in the communitywide total, indicating that the City was falling well short of the 2010 target. Evaluation of individual strategies revealed several useful lessons, particularly the role of greater than expected population growth and the economy on total emissions. In the early 1990s, per capita energy use declined due to federal, state, and local conservation programs. However, this decline reversed between 1995 and 2000 with moderate residential increases and a sharp

Box 8.1
City of Portland and Multnomah County, Oregon, Summary

Plan Titles and Adoption Year: Carbon Dioxide Reduction Strategy (1993), Local Action Plan on Global Warming (2001), and Climate Action Plan (2009)[a]

Plan Author: Portland Bureau of Planning and Sustainability (1993, 2001, and 2009), Mult -
nomah County Sustainability Program (2001 and 2009)

Population: 566,141 (City of Portland), 726,855 (Multnomah County)[b]

GHG Emissions Inventory Year: 2008 (most recent inventory)

GHG Emissions Total: 8,495,319 MTCO$_2$e

GHG Emissions Profile:

— Residential Energy Use	21.0%
— Commercial Energy Use	25.0%
— Industrial Energy Use	15.4%
— Transportation Fuel	38.5%
— Waste Disposal	0.2%

GHG Reduction Target: 40% below 1990 levels by 2030 and 80% below 1990 levels by 2050

Plan Highlights:

— Transit ridership has increased by 75%.

— Bicycling has quintupled with mode share over 10% in many parts of the city.

— City saves $4.2 million annually due to reductions in energy use (~20% of total energy costs and $38 million since 1990)

— 35,000 housing units improved in partnership with utilities and the Energy Trust of Oregon

Source: City of Portland, Oregon. http://www.portlandonline.com/bps/index.cfm?c=49989&.

[a] "Portland Climate Action Now," City of Portland Bureau of Planning and Sustainability, accessed March 4, 2011, http://www.portlandonline.com/bps/index.cfm?c=41896.

[b] "U.S. Census Bureau, 2009 Population Estimates," American FactFinder, U.S. Census Bureau, accessed October 20, 2010, http://factfinder.census.gov.

increase in commercial and industrial emissions. This rise was attributed to a strong economy. Transportation strategies resulted in considerable improvement in transit and bicycle ridership, but community-wide transportation emissions continued to rise. Per capita VMT rose very little in the City of Portland, but increased more than 20% in the larger metropolitan region. This information was critical to future policy development as it indicated areas of needed policy intervention.

The 2001 plan, *Local Action Plan on Global Warming*, made some minor adjustments such as changing the baseline from 1988 to 1990 and revising the goal to a 10% reduction by 2010, closer to the 7% target for the United States under the Kyoto Protocol. Overall, the plan organization and target areas remained the same. The biggest change was the addition of a close partnership with Multnomah County for development and implementation of the plan. Inclusion of the county allowed for explicit recognition of the regional context and formulation of direct action to address regional issues.

The 2001 plan includes a review of implementation success in each sector addressed in the prior plan (1993). The strategies included in the plan are detailed and designed to yield GHG reductions quickly. Each action has a specific, measurable outcome, which makes implementation and tracking progress possible. The 2001 plan also contains one new section focused on education and outreach that includes strategies intended to assure that community members, as well as decision makers, have a clear understanding of climate science, the challenges posed by climate change, and the options for addressing these challenges. During the period following the 2001 plan, Portland and Multnomah County exhibited not only continued per capita GHG reductions but also reductions in overall GHG emissions. By 2005, emissions in the county had been reduced to 1990 levels and were several percent below 1990 levels by 2008. As a result, the City and County are nearly on track to reach the targets set in 2001, though still unlikely to meet the original 1993 goal.

The implementation highlights for the City and County plan during the period between the 2001 and 2009 include the following:

- Transit ridership has increased by 75%.
- Bicycling has quintupled with mode share over 10% in many parts of the city.
- Recycling rates reached 64%.
- The City saves $4.2 million annually due to reductions in energy use (~20% of total energy costs and $38 million since 1990).
- 35,000 housing units have been improved in partnership with utilities and the Energy Trust of Oregon.
- $2.6 billion in annual savings have resulted from reduction in vehicle miles traveled.[2]

The most recent update, the CAP, adopted in 2009, maintains the sectors of focus of the earlier versions with energy efficiency and renewable energy combined into one. Two additional focus areas were added to the 2009 plan: preparing for climate change (e.g., climate change adaptation) and food and agriculture.

The major difference is that the new plan established emissions reduction targets for the combined City and County of 40% below 1990 levels by 2030 and 80% by 2050. The 2050 target will require dramatic emissions reduction. Rather than simply listing this as a lofty but distant goal, the plan spends considerable time detailing a vision for the future and actions to be taken now to achieve this vision. The plan includes both short-term actions that will yield immediate reductions in GHGs and those that will lay the foundation for the vision for 2050. This 2050 vision included a green economy and long-term improvement of the quality of life for all community members. One example of this is the goal of 20-minute neighborhoods "meaning that they (residents) can comfortably fulfill their daily needs within a 20-minute walk from home."[3] Strategies such as this were developed through public outreach events, which were critical given the extent of changes required to meet the 2050 target. The goal of reducing emissions by 80% requires dramatic changes in energy efficiency, energy sources, and travel behavior, such as a reduction from 18.5 to 6.8 passenger miles per day.

Another critical step toward full policy integration is currently occurring in Portland with inclusion of climate principles in the comprehensive plan update.

Lessons Learned

The Portland city and region have engaged in a long, sustained effort in developing and implementing planning policy to address climate change. Over the last two decades, the city and region have achieved some remarkable GHG reductions. Some of the factors identified by the City as keys to this success include a focus on co-benefits, development of partnerships to aid in implementation, and integration of climate goals into all aspects of community policy.

Co-benefits can include improved air quality, human health, economic savings, and greater convenience. These co-benefits not only aid in fulfilling a range of other community goals but also yield unexpected collaborations and garner supporters of the plan. A good example of co-benefits is a strategy to conduct energy-efficiency retrofits in Portland. During the recent economic downturn, few new buildings or homes were being built. The energy retrofit program offers employment opportunities in one of the hardest hit industries, construction. Due to the co-benefit of providing job opportunities, labor unions have begun to actively support the program.

Forging partnerships with community organizations is viewed by City staff as a critical component of successful implementation. Portland has collaborated more closely than many cities on such relationships. They have not only relied on them to aid in facilitating public engagement, but have even given responsibility of implementing some strategies to local organizations. These partnerships provide resources necessary for implementation in the form of funding, material resources, and labor. This allows the City and County to implement more of the identified strategies in a timely manner because it expands the resources available to do so.

The strategies included in a CAP touch on all aspects of community policy. Over time, these principles should be incorporated into the various policy documents that govern a community. It is inefficient and ineffective if the CAP has different strategies than those held in other City policies such as engineering standards, fleet management, land use, transportation, and more. The principles in the various CAPs developed in the region have been incorporated into existing plans as part of standard periodic updates and found their way into the operating procedures of city departments.

When asked what advice they would give a community just beginning a climate planning process, Portland staff identified two basic principles. The first is to start with the easy things. Choose lots of actions that can have immediate impact. Demonstrating effectiveness quickly can build the momentum and gather the support necessary to take on larger, more expensive, and longer-term efforts. The second piece of advice is to learn from others. In the development of the 2009 plan, Portland staff read through the climate action strategies developed by other cities. A city just beginning the CAP process should examine these plans, learn from them, and identify strategies that can be modified or adjusted to maximize local effectiveness.

City of Evanston, Illinois: Empowerment from the Grassroots[4]

To the citizens of the City of Evanston, Illinois, the decision to prepare a CAP was the obvious and right thing to do (box 8.2). The commu-

---

Box 8.2
City of Evanston, Illinois, Summary

Plan Title: *Evanston Climate Action Plan*

Publication or Adoption: November 2008
Plan Author: Community-based task forces organized by the Network for Evanston's Future
City Population: 73,181[a]
GHG Emissions Inventory Year: 2005
GHG Emissions Total: 1.02 MMT CO_2e
GHG Emissions Profile:

— Commercial   58%
— Residential   26%
— Transportation   14%
— Other   2%

GHG Emissions Reduction Target: 7% below 1990 by 2012
Plan Highlights:

— Encourage mixed-use, green, high-performing, transit-oriented development.
— Investigate the feasibility of offshore wind power generation in Lake Michigan.
— Promote local food production, farmer's markets, and food co-ops.
— Establish Evanston's Climate Action Fund, a local carbon offset program.
— Work with the twenty largest local businesses, industrial and institutional energy consumers to establish and meet energy-efficiency and greenhouse gas emissions reduction targets.
— Support efforts to make biodiesel commercially available to residents and businesses.

Source: City of Evanston, Illinois. http://www.cityofevanston.org/pdf/ECAP.pdf.

[a] "2006–2008 American Community Survey 3-Year Estimates," American FactFinder, U.S. Census Bureau, http://factfinder.census.gov.

---

nity has a long history of an active and engaged citizenry who see the issue of sustainability as a key to a better future. Leaders in city hall recognized that waiting for the federal government to solve the problem was neither sufficient nor likely and that local action could be meaningful and effective. In November 2008, the community completed the Evanston CAP; built on a cooperative effort among community organizations, businesses, religious institutions, and government, it showcases what communities can do when they come together to take responsibility for their actions and plan for a better future.

Over the period of a decade, citizens moved from being loosely organized around the idea of creating a sustainable city to creating a visionary and ambitious CAP. Creation of the plan involved organizing these motivated citizens into teams to assist in inventorying the city's GHG emissions, develop potential strategies to reduce these emissions, and compile them into a guiding document that could be successfully implemented. The mayor of Evanston, Elizabeth Tisdahl,[5] described this as a "wonderful process." The "wonder" of Evanston is the community's ability to build coalitions of organizations and volunteers through a truly grassroots effort that was not controlled by city hall.

In the late 1990s, a group of citizens interested in sustainability and the role of the faith community formed the Interreligious Sustainability Project. They held numerous public forums to present and discuss sustainability ideas, and they sought to build a network of interested citizens. This is often referred to as building social capital. The citizens of Evanston weren't organizing to specifically address climate change or any particular environmental issue at all. Instead they were laying the groundwork for future action by educating, inspiring, and networking. This social capital paid dividends several years later with the formation of the Network for Evanston's Future. Although this level of social capital may not be necessary for a successful planning effort, it was perceived by community leaders as inspiring the process and making it more efficient and effective.

With the community coalescing under the Network for Evanston's Future, the city government began the initial steps for climate action planning. In 2006, the Evanston City Council voted unanimously to sign the U.S. Mayors Climate Protection Agreement, and State Representative Julie Hamos committed some of her discretionary funds to support an Office of Sustainability in the mayor's office. With the City on board and the Network for Evanston's Future organizing the volunteer labor, Evanston was positioned to engage in climate action planning. The plan describes the process:

> Rather than hire a consultant or have City staff author a plan for the community, the City embarked on a unique, collaborative partnership with the Network for Evanston's Future, a local sustainability coalition. Nine task forces were established; each with one City [government] and two citizen co-chairs, and the planning process was launched at a community meeting in November 2007 that was attended by more than 130 community members.

Participants were invited to join one of the nine task forces and help develop the recommendations of the Evanston Climate Action Plan (ECAP). In a remarkable display of citizen action, the task forces worked through the winter to research recommendations. A draft ECAP was presented to the community in May 2008 at an Earth Month event attended by more than 300 community members.[6]

The nine task forces covered the following areas:

- Transportation and land use
- Energy efficiency and buildings
- Renewable energy resources
- Waste reduction and recycling
- Food production and transportation
- Forestry, prairie, and carbon offsets
- Policy and research
- Education and engagement
- Communications and public relations

Task force membership was open to anyone in the community who volunteered, which in the case of the Energy Efficiency and Building Task Force, meant about thirty people at first. Although the number dwindled over time, a common problem with volunteer labor, there remained a sufficient core of committed citizens. The task forces were given a blank slate and met monthly for about one year. Often citizens are expected to respond to recommendations generated by government experts. But in Evanston the task forces started with their own brainstorming and defined their own rules for valid ideas. This open, grassroots approach was good for getting numerous creative ideas but also presented a challenge for managing the meetings and information. Ultimately the co-chairs—a local government staff member and a citizen volunteer—played a strong role in managing the process so that it was effective, and the City staff, although primarily there just to represent city interests, helped consolidate ideas into a coherent plan.

The resulting Evanston CAP first explains the consequences of climate change for the region and then explains Evanston's contribution to this problem through an inventory of its GHG emissions. The forecasted consequences of climate change for the Midwest include the following:[7]

- Temperatures in the northern portion of the Midwest are projected to increase by 5 to 10°F by the end of the century.
- Precipitation is projected to increase another 10 to 30% over the region, with much of it coming from heavy and extreme precipitation events.
- Higher temperatures will lead to increased evaporation and lower water levels in the Great Lakes.
- Increased evaporation will also cause soil moisture deficits and more drought-like conditions in much of the region.

Although Evanston on its own cannot counteract climate change and stop or slow these trends, the community knew it had a responsibility to do its part. According to the baseline inventory, the Evanston community emitted 1.02 MMT CO_2e in 2005, mostly due to energy consumption for residential, commercial, and industrial uses.

The plan contains over 200 emissions reduction strategies; "when added together, the strategies have the potential to reduce Evanston's emissions by 245,380 $MTCO_2E$, representing nearly twice the reduction goal of 140,104 $MTCO_2E$."[8] To implement these strategies the community divided responsibility between the city government, as led by the Sustainability Coordinator in the mayor's office, and the community, led by Citizens for a Greener Evanston (CGE), which evolved from the Network for Evanston's Future. As of 2010 the community had made progress on several of the 200+ strategies; for example, they had expanded car-sharing opportunities, added downtown bike racks, launched a local carbon offset program, adopted a citywide green building ordinance, built a green fire station, and improved the energy efficiency of numerous homes, businesses, and public buildings.

Despite its successes, Evanston faces a number of challenges in implementing its plan. A visionary and complex idea to place wind turbines in Lake Michigan has met with skepticism. Infill development using the transit-oriented development principle has met with concerns over the negative impacts of density. Enhancing transit to address regional commuting runs up against the reality of the Chicago region's politics. And of course money is limited. Though, as Mayor Tisdahl reminds those considering climate action planning: "local government is where the action is—it is where things happen."

Lessons Learned

Evanston citizens felt like they could not wait for the state or federal government to take steps to alleviate the climate change problem. Before doing formal planning they built social capital and collaborations, called on organizations already doing good climate action work, and began to build the political will in the community. They then organized and educated themselves and showed that motivated citizens, with support from city hall, can successfully prepare a CAP.

Evanston's impressive success in getting a few hundred citizens to participate in the planning progress was tempered by the reality of having to manage the input and expectations of that many people. They suggest spending time up front developing the mechanism for managing a successful public participation program such as addressing meeting management including rules of participation, communication methods among all the organizations and individuals, and methods for making final decisions.

The Evanston CAP has many strategies that will be implemented by entities within the community such as businesses, nonprofit organizations, and community groups. It is not clear at this point where the resources will come from to implement these strategies or how these entities can be held accountable. When depending on community-based entities, rather than government agencies, to implement strategies there should be significant discussion of these implementation issues.

Evanston found preparing a GHG inventory to be a significant burden that distracted from the planning efforts. They felt that seeking expert assistance on this particular task would free up volunteers to focus on policy and action.

City of Pittsburgh, Pennsylvania: The Power of Partnerships[9]

Like the Evanston plan, the Pittsburgh CAP is the result of a grassroots community effort, rather than a product exclusively of the municipal government (box 8.3). The Pittsburgh climate action planning process was community initiated and catalyzed by several organizations. In 2006, the Green Building Alliance, with then mayor Bob O'Connor, convened the Green Government Task Force of Pittsburgh to begin discussions of addressing sustainability and climate change. At the same time, Green Building Alliance partnered with Carnegie Mellon University students and faculty to develop the City's first GHG emissions

Box 8.3
City of Pittsburgh, Pennsylvania, Summary

Plan Title: *Pittsburgh Climate Action Plan, Version 1.0*
Publication or Adoption: June 2008
Plan Author: Pittsburgh Green Government Task Force
City Population: 295,988[a]
GHG Emissions Inventory Year: 2003
GHG Emissions Total: 6.6 MMT CO_2e
GHG Emissions Profile:
— Commercial   56%
— Residential   18%
— Industrial   6%
— Transportation   20%
GHG Emissions Reduction Target: 20% below 2003 by 2023
Plan Highlights:
— Plant over 20,000 trees throughout the city in the next five years.
— Convert traffic signals to LED lights saving nearly $70,000 annually for a 15% conversion.
— Create a full-time city sustainability coordinator.
— Require LEED (green building) certification for all municipal buildings.

Source: City of Pittsburgh, Pennsylvania. http://www.pittsburghclimate.org/index.htm.

[a] "2006–2008 American Community Survey 3-Year Estimates," American FactFinder, U.S. Census Bureau, http://factfinder.census.gov.

inventory. In addition the Surdna Foundation, a private grant-making foundation, helped bring in and fund the nonprofit Clean Air–Cool Planet to provide technical expertise and write the plan. The city government was a partner in the process but did not take the lead or oversight role. In Pittsburgh, climate action planning was a stakeholder-based process to produce a plan with investment from all sectors.

The Pittsburgh *Greenhouse Gas Emissions Inventory* was completed in 2006 by a student and faculty research team from the Heinz School at Carnegie Mellon University. The baseline year for analysis was 2003 because this was the earliest year that complete data were available. The team used ICLEI's Cities for Climate Protection software to organize and analyze GHG emissions data. The City emitted about 6.6 million tons of CO_2e GHGs in 2003. Of these emissions, 56% were attributable to the commercial sector, 18% to the residential sector, 6% to the industrial sector, and 20% to the transportation sector. Another way to look at the emissions is by fuel type; in this case 72% of the city's emissions were from burning fossil fuels for electricity. Because the city's population is not growing—and perhaps is shrinking—the business-as-usual (BAU) forecast shows about a 2% reduction by 2015. The plan established a GHG emissions reduction target of 20% below 2003 by 2023 that was based on a review of peer communities and feasibility discussions of the task force.

The plan contains recommendations for government, business and industry, the community at large, and higher-education institutions. These recommended measures describe the action to be taken, the cost and savings, the time frame, and the resulting GHG reductions. The following is a list of some of the more innovative measures:

- Plant over 20,000 trees throughout the city in the next five years.
- Create a full-time city sustainability coordinator.
- Require Leadership in Energy and Environmental Design certification (green building) for all municipal buildings.

To fund implementation of the plan, Pittsburgh is using a combination of funding sources, including state and federal grants. In 2008, Mayor Ravenstahl created a Green Trust Fund of $100,000. This seed money came from the savings that resulted from beginning to purchase electricity through a reverse auction. As the City implements energy-saving projects, it can roll the savings back into the trust fund to fund continued energy improvements.

By summer 2007 the planning process was well under way. The Green Government Task Force, a group of twenty-nine community leaders representing a wide array of interests, established five working groups in the areas of business, community, higher education, municipal, and communications. Visioning meetings were held in numerous locations around the city and citizens were asked for their ideas. These ideas were collected and disseminated to the working groups. Clean Air–Cool Planet and Green Building Alliance worked with the working groups to develop sections of the plan. The Green Government Task

Force reviewed the draft plan over several meetings. In August 2008, the plan was adopted as a guiding document by the city council and Mayor Luke Ravenstahl. According to the plan, the community undertook this effort "to reduce the impacts of local and global climate change, improve the local environment and the local economy, and enhance Pittsburgh's reputation as an environmentally progressive city."[10] In this statement, Pittsburgh is identifying numerous benefits of developing a CAP.

To implement the plan, the Pittsburgh Climate Initiative was formed as an umbrella for the following organizations (with roles shown):[11]

- *Management*: Green Building Alliance—Pittsburgh Climate Initiative facilitator and convener, manages measurement and verification of Pittsburgh's reduction goal, which is to reduce GHG emissions 20% below 2003 levels by 2023.
- *Government*: City of Pittsburgh and Allegheny County governments—working to reduce GHG emissions from City and County operations facilities.
  - Pennsylvania Environmental Council—working directly with Allegheny County on GHG inventories and action plan for county authorities.
- *Community*: Citizens for Pennsylvania's Future—community outreach leader of The Black and Gold City Goes Green initiative, which tracks individual actions to reduce GHG emissions.
- *Business*: Sustainable Pittsburgh—Business Climate Coalition convener.
- *Higher education*: the Higher Education Climate Consortium (HECC) actively engages all Pittsburgh region colleges and universities to collaborate, share information, and set goals so that HECC organizations can align with Pittsburgh's overall GHG reduction goal. At their request, HECC is convened by Green Building Alliance.

Pittsburgh has chosen a decentralized approach for implementing its CAP. The Pittsburgh Climate Initiative partners meet once per month to coordinate on implementation of the plan.

At the City of Pittsburgh, responsible for the municipal section of the plan, the mayor created a sustainability coordinator position (later ratified by the city council) and a Sustainability Commission to oversee implementation of the plan. In fall 2010, the City completed its second GHG inventory, using baseline 2008 data. As of late 2010 the City had completed or was fully in progress on eighteen of the twenty-three municipal emissions reduction measures.

Pittsburgh has begun the process of updating the plan with a focus on developing more specific emissions reduction strategies and measurable targets. For example, the City will be considering a specific renewable energy target for supply of city energy needs. In addition, the update will tackle some of the more challenging areas for emissions reductions that were not addressed in the first plan.

Lessons Learned

In Pittsburgh, strong leadership from the mayor and the formation of a green ribbon–style task force of prominent community leaders was an important first step in creating a CAP. These community leaders represented a diverse set of important sectors in the Pittsburgh economy. Forming this type of coalition of local partners built broad support for the plan and ultimately improved the quality of the plan.

With the large number of participants and partners involved in the Pittsburgh planning process, establishing a clear organization with assigned roles and responsibilities early in the process proved to be critical to an efficient planning process. Perhaps most importantly, the role of city government was clarified in the process since it was acting as one of the partners and not as the lead planning entity.

Finally, Pittsburgh took a unique approach to implementation by assigning it out to different groups rather than housing it under one nonprofit agency or City department as is usually the case. For this, getting the right people in the room—those with working knowledge in their sectors—was important for developing and implementing emissions reduction strategies.

City of San Carlos, California: Integrating CAPs and General Plans[12]

To understand some of San Carlos's choices, it is important to recognize that the California context for climate action planning is different from other states. Through a series of governor's executive orders, legislative acts, and court rulings the state has created compelling incentives for preparing a local CAP. State law directs that GHG emissions must be considered in most planning and development decisions, and reduction

strategies should be implemented as feasible. Most local governments
are finding that the clearest way to meet these requirements is to create
a CAP rather than deal with them incrementally or through other
policy instruments. Most planners, attorneys, and their professional associations are recommending the same, thus making local climate action
planning in California as close to a mandate as exists in the United
States.

   In 2008, the San Carlos City Council adopted the "City of San
Carlos Climate Protection Letter" (box 8.4). The letter committed the
City to act to protect the climate but it also called on the federal and
state governments to do their part. Prior to this the City had already
shown a commitment to addressing climate change by being a charter
member of a regional effort called the Silicon Valley Climate Protection
Initiative and by developing the Community Solar Discount Program
in partnership with Solar City and San Carlos Green. The City of San
Carlos is an example of a community that had already made significant
commitments to GHG reductions, energy efficiency, and overall sustainability before starting the climate action planning process; it was not just
responding to the state's new push for climate action planning.

   The City followed these actions by adopting the *San Carlos 2030
General Plan and City of San Carlos Climate Action Plan* (CAP) in October 2009. The planning process was led by the City's Community Development Department with support from a three-member Climate
Action Plan Subcommittee of the General Plan Advisory Committee
and a consulting firm specializing in climate action and sustainability.
The process took about two years and included numerous public workshops as well as participation by several community and regional public
(and quasi-public) agencies involved in transit, housing, environment,
public health, air quality, law, and economic development. The City
community development director described the process: "Our community put forth a tremendous and thoughtful two-year effort on these
important plans for our future. It was a collaborative effort involving San
Carlos youth, residents and businesses. The process also involved many
partnerships and resources throughout the region."[13]

   The most innovative aspect of the CAP was that it was developed
in parallel with the General Plan and integrated into the City's General
Plan. This was done by linking land use and transportation policies and
programs in the General Plan to measures to reduce GHG emissions in

Box 8.4
City of San Carlos, California, Summary

Plan Title: *City of San Carlos Climate Action Plan*

Publication or Adoption: October 2009

Plan Author: City of San Carlos Planning Department and the General Plan Advisory Committee (GPAC) Climate Action Plan Subcommittee, with consulting services from PMC, Inc.

City Population: 28,652[a]

GHG Emissions Inventory Year: 2005

GHG Emissions Total: 267,237 MT CO_2e

GHG Emissions Profile:

— Commercial and Industrial   20%
— Residential   18%
— Transportation   56%
— Waste   5%

GHG Emissions Reduction Target: 15% below 2005 levels by 2020, and 35% below 2005 levels by 2030

Plan Highlights:

— Integrated land use and transportation policies and programs in the Environmental Management Element of the General Plan with measures to reduce GHG emissions in the CAP.

— Increase overall solid waste diversion to recycling by at least 1% per year.

— Provide for increased albedo (reflectivity) of urban surfaces including roads, driveways, sidewalks, and roofs in order to minimize the urban heat island effect.

Source: City of San Carlos, California. http://www.cityofsancarlos.org/generalplanupdate/whats_new_/climate_action_plan___adopted.asp.

[a] "2006–2008 American Community Survey 3-Year Estimates," American FactFinder, U.S. Census Bureau, http://factfinder.census.gov.

the CAP. In addition, all of the strategies identified in the General Plan and CAP are integrated into a single implementation plan that addresses phasing, method of implementation, the entity responsible for implementation, and cost. This integration was facilitated by forming the Climate Plan Committee as a subcommittee of the General Plan Advisory Committee. This integration of climate action planning and general or comprehensive planning has little precedent but will likely become a more common practice.

In addition to the general plan integration, the City of San Carlos CAP is innovative in the way it addresses establishing and achieving the GHG reduction target. The CAP's goals for GHG reductions, 15% below 2005 by 2020 and 35% below 2005 by 2030, were established consistent with the state's California Global Warming Solutions Act (AB 32). The CAP established twenty-three strategies for reducing GHG emissions and achieving the target. Prior to setting the City's reduction strategies, the City forecast emissions to account for state initiatives that require utilities to increase the amount of renewable energy in their portfolios and changes aimed at reducing GHG emissions from automobiles. The City of San Carlos has taken the prudent measure of accounting for external change in order to set a more realistic level of local responsibility for GHG reductions. The CAP also addresses climate adaptation and identifies potential adaptation strategies.

In recognition of the plan being "practical and highly implementable," the CAP won awards in 2009 and 2010 from the Northern California Chapter of the American Planning Association.[14] The San Carlos mayor, Randy Royce, had this to say:[15]

> The *San Carlos General Plan 2030* represents a significant upgrade to the City's planning process. The integration of the *2030 General Plan* and the *Climate Action Plan* is a progressive step in linking land use with citywide environmental sustainability. These links have drawn attention to the City's approach to the General Plan and *Climate Action Plan* from cities and agencies across the region. This award further acknowledges that San Carlos is moving in the right direction by focusing on Smart Growth goals such as walkability, improving connectivity within the community and striking a balance between housing, jobs and the El Camino Real transit corridor.

Lessons Learned
From the beginning San Carlos planned to integrate the CAP with the community's General Plan. This was a much better strategy than preparing the CAP and then trying to make it fit into the General Plan. This integrated approach ensures full coordination from the goal level down to the specific program or action level. This integration was solidified through a unified Implementation Program for both plans. San

Carlos's actions may be an indicator of a trend toward greater integration of climate issues across the variety of community plans.

Leadership from the community took some of the burden off of the City and created a broad base of support for the plan. Residents formed a nonprofit task force called San Carlos Green and worked with City staff to facilitate community outreach and involvement. The Chamber of Commerce created a green business task force which promoted green business practices through a column in its newsletter, developed a green business tradeshow, and helped support the countywide green business certification program.

San Carlos made the decision to supplement City staff who were leading the plan development with a hired consultant specializing in climate and sustainability services (PMC, Inc.). The consultant provided assistance in all phases of the plan process but was most helpful in two areas. The first was developing the technically demanding GHG emissions inventory and forecasts. Since California communities are now basing some land use decisions on information from GHG inventories these inventories must be technically robust and legally defensible. The second was to assist with public participation in the planning process. San Carlos wanted an extensive public outreach and participation program beyond what staff would be able to manage and deliver on their own.

Miami–Dade County, Florida: Striving for Sustainability[16]

Miami–Dade County, Florida, is one of the twelve original members of the ICLEI Cities for Climate Protection Campaign founded in 1990 (box 8.5). In 1993 the County became one of the first communities in the world to create a CAP. Called *A Long Term CO$_2$ Reduction Plan for Metropolitan Miami-Dade County, Florida,* it set a GHG emissions reduction target of 20% below 1988 levels by 2005 and identified thirteen areas for emissions reduction. Also, it notably called on the state and federal government to adopt measures to improve vehicle gas mileage and energy conservation.

In 2006 the County updated the plan and reported that as a direct result of the implementation of the plan, the County's CO$_2$ emissions reductions averaged 2.5 million tons per year.[17] However, Miami-Dade County's 27% population growth over the thirteen years of the planning time frame resulted in an overall increase in emissions. Although the

Box 8.5
Miami-Dade County, Florida, Summary

Plan Title: *GreenPrint: Our Design for a Sustainable Future*

Publication or Adoption: December 2010

Plan Author: Miami-Dade County

County Population: 2,457,044[a]

GHG Emissions Inventory Year: 2005

GHG Emissions Total: 30.7 MMT CO_2e

GHG Emissions Profile:

— Transportation   43%

— Commercial   25%

— Residential   25%

— Waste   4%

— Industrial   3%

GHG Emissions Reduction Target: 20% below 2008 by 2020.

Plan Highlights:

— Be green government role models and leaders in energy, fuel, and water efficiency.

— Expand alternative fuel (biodiesel/waste-based biodiesel) and renewable energy industries.

— Protect environmental and other lands important for ecosystem and community resilience.

— Increase transit ridership including linking the Metrorail to Miami International Airport (MIA).

— Make fresh, local, organic food available through grocers, farmer's markets, and community gardens.

Source: Miami-Dade County, Florida. http://www.miamidade.gov/greenprint/.

[a] "2005–2009 American Community Survey 5-Year Estimates," American FactFinder, U.S. Census Bureau, http://factfinder.census.gov.

plan kept the emissions lower than they might have been otherwise, the County realized the challenge of reducing emissions in a fast-growing community. The County also cited the failure of the state and federal governments to take more aggressive action as part of the problem.

Also in 2006, the County created the Miami-Dade County Climate Change Advisory Task Force (CCATF), made up mostly of technical experts in the climate change field, to advise the mayor (the County is a municipal county) and the Board of County Commissioners. The task force produced annual reports and investigated new strategies for re-ducing emissions. Most importantly, the task force kept the County mov-ing forward with climate action planning that would lead to the next evolution of progressive action on energy and environmental issues.

In 2009, ICLEI chose Miami-Dade County as one of only three communities for pilot testing its new Sustainability Planning Toolkit. The County saw this as an opportunity to revisit the CO_2 reduction plan, broaden its scope, and use it to coordinate a variety of ongoing and new activities across departments and the community. The result of this process was the *GreenPrint: Our Design for a Sustainable Future* (*GreenPrint*) sustainability plan released in December 2010. Within *GreenPrint* is a chapter called the "Climate Change Action Plan," which has a GHG emissions inventory, reduction targets and strategies, and its own distinct identity from the larger sustainability plan.

To develop *GreenPrint*, the mayor created the Sustainability Advi-sory Board made up of community leaders from the business, higher education, and nonprofit sectors. The mayor also established a sustain-ability director position to oversee the planning and implementation process. The strategy of Sustainability Director Susanne M. Torriente was to elicit input and participation from everyone in the community and treat them as partners in the sustainability effort. Ms. Torriente re-ported that the process was a positive one and that the sustainability theme, rather than climate change, helped when dealing with skeptics. Many saw the proposals in the plan as "common sense" actions that would improve the "quality of life" in the community, save the com-munity money due to energy efficiency, or provide other benefits. In a sense, emissions reduction and climate adaptation became co-benefits rather than the usual primary benefits.

Implementation, which is just beginning, is anticipated to be daunting due to the sheer number of initiatives. Answering questions from department heads about how to get started, where to find funding, and how to track progress has been first on the list of things to do. In addition, the County is reaching out to community partners to support implementation efforts, particularly for in-kind services. One of the an-ticipated keys to success is the County's decision to tie the plan to the budget process by developing a sustainability scorecard that is consistent with the County management scorecard for measuring progress (key component of allocating resources). In addition, sustainability is being integrated into regular County operations so that funding is not seen as competing, and moreover, showing departments how to save money. For example, a "cool roof" installation at one of the County libraries has saved $1,000 per month on electricity costs.

Miami-Dade County has a Comprehensive Development Master Plan (CDMP) that "expresses the County's general objectives and policies addressing where and how it intends development or conservation of land and natural resources will occur during the next ten to twenty years."[18] The County's Planning and Zoning Department was involved throughout the whole planning process to ensure that *GreenPrint* and the CDMP were consistent with each other. The goal is that *GreenPrint* can serve as a focusing and implementing tool for the long-range goals in the CDMP. In addition, the graphics-heavy, user-friendly *GreenPrint* is a great public relations tool for the more technical, regulatory CDMP; it says many of the things the CDMP says in a much more accessible way for the public and county officials.

Miami-Dade County is not finished planning. The County is now building on the success of the ICLEI Sustainability Planning Toolkit by participating as a pilot community for ICLEI's Climate Resilient Communities (CRC) Program to develop a climate adaptation plan. In addition, Miami-Dade County has joined with Broward, Palm Beach, and Monroe Counties to form the Southeast Florida Regional Climate Change Compact with the purpose of preparing a regional CAP.

Lesson Learned

According to Miami-Dade County, one of the most important lessons learned was that an issue could be studied for years, but that this should not delay getting started. To get started, the County secured strong and vocal support from the mayor and board, created a sustainability director position, and staffed it with someone good at planning, organization, and task management. This was critical for the success of climate action planning in a complex urban county with so many regional assets.

Building peer relationships and community partnerships proved to be key for Miami-Dade County. The County reached out to numerous communities that had completed sustainability plans and CAPs to learn about their experiences and get advice. In particular the County received great help from the City of New York, which has recently completed its *PlaNYC* sustainability and climate action plan. In addition the County received significant technical support from the National Oceanic and Atmospheric Administration Coastal Services Center and had the strong backing of the local tourism board, which represents a major part of the south Florida economy.

Many communities struggle with how to get people interested in and supportive of climate action planning. Although the impacts of climate change seem like they would constitute an important motivator, the County found that this tended to scare or put people off. Instead, the County focused on energy efficiency and quality of life benefits and found the public and officials much more receptive. Figuring out what is important to the community when starting the process is a key to success.

Finally, the County found the process to be very iterative. In other words, as staff moved through the process and tried things or learned new information they sometimes had to back up and do things over. For example, some of the ICLEI tools proved to be too complicated to work with so participants in the process had to revise the tools and start again. The County advised participants not to be afraid to jump in and get to work.

City of Homer, Alaska: Big Plans in Small Places[19]

The City of Homer, home to just under 6,000 residents, is located on Kachemak Bay in south-central Alaska. The Homer CAP[20] represents a proactive approach to climate action planning taken by a small community, demonstrating that climate action planning can and should occur in all city types, locations, and sizes (box 8.6). The plan was developed by the local Global Warming Task Force in collaboration with ICLEI– Local Governments for Sustainability (ICLEI). The task force coordinated community outreach efforts, data collection, and strategy development. A critical lesson learned from Homer is that a CAP is only as effective as the implementation that occurs following plan adoption. Homer has not only begun implementing individual policies in the plan but has laid a foundation that will support long-term effectiveness. These measures include integration of CAP strategies into the comprehensive plan and economic development plan, establishment of funding mechanisms to support strategy implementation, and organization of a process for monitoring data to track implementation.

Box 8.6
City of Homer, Alaska, Summary

City of Homer Climate Action Plan

Publication or Adoption: December 2007

Plan Author: Community-Based Global Warming Task Force and City staff

City Population: 5,667[a]

GHG Emissions Inventory Year: 2006

GHG Emissions Total: 135,621 Tons CO_2e, backcast to year 2000: 98,123 tons CO_2e

GHG Emissions Profile:

— Commercial   36%

— Residential   24%

— Transportation   21%

— Marine   17%

— Waste   2%

GHG Emissions Reduction Target: 12% below 2000 by 2012 and 20% below 2000 by 2020

Plan Highlights:

— Clear articulation of reasons for Homer to engage in climate planning

— Goals to pursue feasibility assessment of renewable energy generation

— Establishment of ongoing, detailed data collection to track implementation

— A diversity of funding mechanisms identified to support implementation

— Identification of adaptation measures that recognize the need for close ties to science entities and a systemwide approach

— Integration of climate action planning principles in other city policy, including the comprehensive plan and economic development plan

Source: City of Homer, Alaska. http://www.cityofhomer-ak.gov/citycouncil/climate-action-plan.

[a] "2005–2009 American Community Survey 5-Year Estimates," American FactFinder, U.S. Census Bureau, http://factfinder.census.gov.

In September 2006, Mayor James C. Hornaday attended a conference on climate change. This event sparked a series of actions that culminated in the adoption of a CAP a year later. Local action to initiate the process began with establishment of the Global Warming Task Force followed by joining ICLEI. The task force was formed by soliciting applications from interested community members who were subsequently approved by the city council. Staff support was provided by the city manager's special projects coordinator, who provided informational and organizational resources. The climate planning process began with an evaluation of CAP development and strategies employed by other

U.S. cities and local public outreach and education. After this evaluation, CAP development followed the standard sequence of conducting an emissions inventory and then developing GHG reduction strategies and adaptation measures.

Adoption of the completed CAP was the next hurdle. In the end, it was approved by a 3–3 vote, with the mayor breaking the tie. CAP advocates credit the outpouring of community support along with the political savvy of the task force chair, who was a former city council member, in achieving passage of the plan.

The CAP itself spends considerable time explaining the reasons for Homer to engage in local climate policy development, including the particular threats posed by climate change to northern-latitude locations like Homer, the local public policy opportunities, the chance to serve as a leader in the State of Alaska, and the ethical basis for taking action. Following a summary of climate science and the local GHG emissions inventory, the CAP establishes reduction measures in the following sectors: energy management, transportation, purchasing and waste reduction, and outreach and advocacy. Reduction strategies of note include specific direction for updating the City's comprehensive plan, renewable energy pilot programs, aggressive energy-efficiency contract conditions for city facilities, establishment of a bike-library, nonmotorized transportation and trails plan implementation, and detailed regional advocacy goals. The reduction measures chapter is followed by chapters detailing adaptation measures and implementation.

The adaptation chapter demonstrates a clear understanding of the dynamic context in which climate adaptation must occur. It begins by addressing the challenges presented by the ongoing evolution of climate science by tightening the feedback loop between science entities and city policies though a relationship with the University of Alaska–Fairbanks Center for Climate Assessment and Policy. In addition, the listed policies demonstrate recognition that the adaptation strategy must address not only structural threats and emergency preparedness but also the many other sectors of city function that may be impacted. There are strategies that identify the need to assure a resilient economic sector, improve communication networks, and support a well-informed population through education.

Following identification of reduction and adaptation measures is a chapter detailing a wide variety of funding mechanisms. While not binding, the range of potential funding sources identified in the plan allows for flexibility to assemble varying combinations of revenue to support the strategies identified. These implementation tools include parking fees, carbon tax, state and federal grants, offset programs, and the savings from energy efficiency upgrades. These funds are to be held in a newly established Sustainability Fund. The plan continues by providing a list of specific projects and programs that the fund would be used to support.

Lessons Learned

Over the three short years since Homer's CAP was adopted a tremendous amount has been accomplished. The biggest lesson to be learned from the experience of climate planning in Homer is the importance of implementation. Achievements include establishment of ongoing monitoring and a protocol for reporting, funding mechanisms, educational programs, and integration of climate planning principles into many existing City plans. These accomplishments can be attributed to the practical nature of the plan, continued involvement and advocacy by community members, and the political will of the City leadership.

Many of the indicators defined as part of a city's CAP are already collected as part of standard operations. The challenge for many cities is often that these data are housed in a variety of different offices, and there is no person or department that collects all relevant data and evaluates implementation success. The *City of Homer Climate Action Plan Implementation Project Final Report*[21] provides detailed energy information to serve as the foundation for implementation of strategies in all city-owned facilities regardless of department or function. This report also established a protocol and central location for continued monitoring of implementation progress. The development of this resource will allow for progress reports and plan updates to be more easily assembled and informed decisions to be made. A Revolving Energy Fund was established to provide revenue for energy efficiency projects in city facilities. Already this fund has been used to support detailed energy audits and efficiency upgrades at sixteen facilities.[22]

A resource that has received attention from many other jurisdictions, most of them outside the state of Alaska, is a guidebook developed for City employees titled "Money, Energy, and Sustainability."[23] This guidebook was the first of its kind in the United States and provides guidelines for reducing energy use and waste production for all City operations. A strategy such as the guidebook is a critical complement to the facilities upgrades and monitoring described earlier. By engaging and educating the staff that operate and work in these facilities, the guidebook increases the likelihood of reaching and exceeding the emissions reduction targets for government operations.

Similar to achievements in government operations, progress has been made in implementation of community-wide strategies. One of the most impressive steps taken in implementation was the revision of the comprehensive plan adopted in 2010 to include climate planning principles, primarily housed in a new chapter titled "Energy."[24] Integrating the CAP goals into the comprehensive plan allows for consistency in City policy as well as improves the likelihood that emissions reduction and climate adaptation are integrated throughout City governance. For example, the 2011–2016 *Capital Improvement Plan*[25] includes projects focused on facility upgrade and renewable energy. Renewable energy goals have also been implemented through approval of a new ordinance allowing wind turbines on private property. Transportation and land use strategies have been pursued through integration of smart growth principles and enhanced walkability.

Climate adaptation has been integrated into existing plans as well. In February 2011, an economic development plan[26] was adopted that is consistent with the revised comprehensive plan and addresses many of the challenges that climate change may present to the city. The plan includes goals such as food security, renewable energy generation, responsible fisheries management, and local smart growth principles, including affordable housing. The adaptation chapter of the Homer CAP focused on development of a resilient economic sector, including housing, energy, and local knowledge. The economic plan further articulates these goals and demonstrates one more way in which Homer has taken the critical step of integrating emissions reduction and adaptation throughout city policy.

> One of the phrases heard during public testimony in support of the Climate Action Plan is that implementation of the plan is not only the right thing to do, it is also the smart thing to do. Most of the measures recommended in this plan would be prudent even if climate change did not exist.
>
> Mayor James C. Hornaday, City of Homer, Alaska

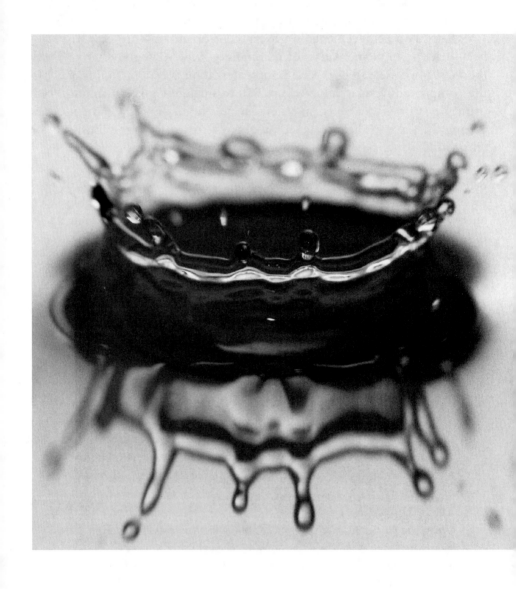

Chapter 9

―――――――― ✖ ――――――――

Time to Take Action

As we continue to tackle our environmental challenges, it's clear that change won't come from Washington alone. It will come from Americans across the country who take steps in their own homes and their own communities to make that change happen.

U.S. President Barack Obama in a speech on
"A New Foundation for Energy and the Environment"
http://www.whitehouse.gov/issues/
energy-and-environment/new-foundation

Strategic local plans focusing on the reduction of greenhouse gas (GHG) emissions and increasing a community's resilience in the face of unavoidable climate impacts are being pursued nationwide. The development of these climate action planning strategies, which are either included in a stand-alone plan or incorporated into comprehensive land use plans or sustainability plans, is likely to continue into the foreseeable future. Development of these plans represents a unique opportunity for communities not only to contribute to solving a global problem but to position themselves to thrive well into the future. Climate action planning should be seen as a chance for communities to control their own destiny in the face of shifting conditions, and to act as leaders in the formation of effective, innovative climate policy.

The Case for Immediate Action

We humans have been contributing GHG emissions into the atmosphere for generations. Communities may ask: Why take action now?

Why not wait until we know more? We offer three reasons for communities to take immediate action to reduce their GHG emissions and prepare for the local impacts of climate change.

The Longer We Wait the Harder It Will Get

The impacts of the accumulation of GHGs in the atmosphere have only been recognized as tied to climate change in the last few decades partly because the concentrations have reached levels where the outcomes are more directly observable. Just as it has taken many decades to create the problem, the solution will require prolonged effort. Climate action planning seeks to reduce emissions that are contributing to the problem and reduce vulnerability to those impacts that are unavoidable.

The benefits from emissions reductions will not be felt for many decades. This alone should be motivation to take action sooner rather than later; however, it can also be viewed as an excuse to delay action. In this case, communities should be concerned about GHG reduction expectations based on international treaty or federal or state mandate. Many of these, such as the Kyoto Protocol or the California Global Warming Solutions Act, have established emissions reduction targets. The longer a community waits to take action the more aggressive, and often expensive, these actions will need to be to meet stated goals. In the future these goals may become mandates that require local action. Cities that take action now are best positioned to satisfy future policy requirements.

In the case of adaptation, the motivation to take immediate action is much more urgent. Many communities face projected climate change impacts that could have far-reaching consequences for a city's infrastructure, economic base, and public safety. What should prompt immediate action is the fact that addressing some of these changes will take considerable time and investment. If climate change impacts potentially place infrastructure such as airports, marine ports, wastewater treatment plants, or major roadways at risk, the resulting damage can be costly in terms of both loss of life and financial loss. For example, if a city's wastewater treatment facility is located near a shoreline, sea level rise or flooding may place the physical structure at risk. If the facility is viewed as so vulnerable that it should be moved, a community would have to identify a new site, obtain funding, build the new facility, and reroute sewer infrastructure to reach the new facility. These steps will all take time and money. A city facing projected climate change impacts cannot afford to delay planning.

Communities Can Achieve Long-Term Success

Acting now to develop GHG emissions reduction and climate adaptation strategies allows communities to control their own destiny. Climate action plans have the power not only to reduce vulnerability to the hazards associated with climate change but to position a city to thrive economically, environmentally, and socially well into the future. The needs to reduce GHG emissions and adapt to unavoidable consequences are likely to be considerations in policy development for many decades. These goals are compatible with the many other goals of local governments such as housing, environmental protection, and economic development. The development of a comprehensive, integrated climate action plan presents an opportunity for communities to take measures that can meet a range of local needs now and well into the future.

The emission of GHGs is a result of how our energy is produced and the efficiency with which it is used; the manner in which people move around a community; the products purchased and the manner in which they were made; and the methods of solid waste disposal. These choices are influenced by local urban form, climate, culture, economic conditions, values, employment base, and a host of other local characteristics. While the overarching goals of climate policy may be shared, the most effective and lasting climate planning strategies will acknowledge and build on local context. Accounting for the local environmental, economic, political, and social setting in the development of a climate action plan provides a great opportunity to incorporate strategies that not only meet the global needs of climate change but situate a community for long-term livability. For example, a shift in transportation mode share from single-occupant vehicles to bus, bike, and pedestrian travel to reduce GHG emissions is also likely to improve public health due to increased physical activity and improved air quality. Improved energy efficiency in homes reduces the utility bills of residents. Strategies that seek to foster a green business community are likely not only to reduce the emissions associated with the commercial and industrial sector but also to create an economic stimulus and foster job creation.

Sustained collaboration on climate action planning strategies can foster community support and security. Community members experience the outcomes of these actions, from safer streets to great economic stability. Over time, these actions yield a slow shift in local culture and understanding. The implementation of climate planning strategies can yield an improved global condition as well as a community that is vital, livable for all residents, and economically resilient. Climate action planning becomes, simply, *good community planning.*

Communities Are Positioned to Innovate and Lead

The federal government has struggled to formulate and pass climate change legislation. Even when countries or international organizations succeed, the policies represent compromise. This is to be expected given the diversity of interests and issues at large spatial scales; actions taken at these scales are often broad in scope and less able to be experimental or innovative. Local communities are the entities best positioned to innovate in the realm of climate action planning.

A community that makes a long-term commitment to climate planning goals should treat strategy development as an adaptive process. Some strategies may not work, but others will prove effective. Each community represents a unique local setting from the biophysical to the sociopolitical. Strategy development built from a laundry list of actions used elsewhere will only be successful if adapted to the specific challenges and opportunities presented by any given community. Local governments have the opportunity to be incubators for new climate action planning strategies that can be tailored or adjusted to meet the needs of others.

Climate action planning is a new enough area of policy development that innovative ideas and approaches are critical to assure widespread success. The long-term nature of climate planning efforts gives communities room to experiment. However, experimentation requires a firm commitment to monitoring. The feedback loop provided by monitoring of GHG emissions and implementation of reduction measures allows for adjustments to be made and areas of success or failure to be identified.

Many areas of climate action planning rely on voluntary behavior such as the choice to drive or walk or the choice of housing type. Climate action plan strategies can render some behaviors inconvenient or expensive and others easier or cost-effective, but these behaviors cannot be directly addressed through policy. Community culture and values change with time. Engaging in a climate action planning process represents a long-term commitment. As local culture changes, there are strategies that may have been ineffective in the past that prove effective in the present. The implementation of strategies and community acceptance or support of climate policy evolve together. Demonstrated or observed policy effectiveness can lead to shifts in community views. The case study of Portland and Multnomah County in Oregon illustrates a slow evolution in transportation choices that occurred as the government focused on providing greater opportunity and convenience with an expansion of transit, bike infrastructure, and pedestrian safety. As residents began to use these services and opportunities, the societal view of them began to shift.

Over time, sustained implementation and monitoring of climate action planning strategies will likely begin to inform the policies that govern many areas of community operation. In the long term, climate planning goals should be integrated into all areas of government. This normalization of policy would result in the goals of minimizing GHG emissions and exposure to hazards associated with climate change becoming standard considerations in all decision-making processes.

Who and for Whom?

Anyone who asks the question, who should do climate action planning for our community? should consider themselves a candidate for the answer. As shown in the case studies in chapter 8, climate action planning can be done in city hall, it can be done by nonprofits in the community, and it can be done by a dedicated and informed citizenry that comes together for this specific purpose. Those motivated to do climate action planning in their community should immediately begin to make connections and develop partnerships as described in chapter 2. Any community can do climate action planning, and should. Moreover, communities can even begin reducing emissions immediately while the climate action planning process gets started (box 9.1).

Box 9.1
Ten Things Your Community Should Do Right Now
(Even Before Starting the Climate Action Plan)

1.   Switch to energy efficient lighting such as CFLs and LEDs.
2.   Upgrade insulation in older residences, businesses, and government buildings.
3.   Install solar panels where feasible.
4.   Purchase high-fuel-efficiency and clean fuel vehicles.
5.   Start or enhance your recycling program.
6.   Provide and promote opportunities for transit, bicycling, and walking.
7.   Use local, sustainably produced products, especially food.
8.   Purchase renewable energy (if available).
9.   Conserve water through retrofit of fixtures, low-water landscaping, and rainwater catchment.
10. Plant trees.

We dedicated this book to our young nephews and nieces. Right now they know very little of the world around them. They don't know that climate change could compromise the social, cultural, and economic integrity of the communities in which they live. Right now they are happy, playful, inquisitive, fulfilled, and loving. Through the actions described in this book to deal with climate change now rather than later, we hope to keep them that way.

Appendix A:
Climate Science

The effort to organize and interpret the hundreds of scientific studies about various aspects of climate change is being led internationally by the Intergovernmental Panel on Climate Change (IPCC) and in the United States by the U.S. Global Change Research Program (USGCRP). Communities can use the data and summary reports from these entities to establish the scientific basis for taking local action. The IPCC was "established by the United Nations Environment Programme (UNEP) and the World Meteorological Organization (WMO) to provide the world with a clear scientific view on the current state of climate change and its potential environmental and socio-economic consequences."[1] The IPCC is made up of thousands of scientists from around the world who review and assess the latest science and publish a report every several years, most recently in 2007. The USGCRP was established by the Global Change Research Act of 1990 (P.L. 101-606), which called for "a comprehensive and integrated United States research program which will assist the Nation and the world to understand, assess, predict, and respond to human-induced and natural processes of global change."[2] It is composed of thirteen federal agencies and departments[3] and publishes an annual report titled *Our Changing Climate* as well as numerous other reports, including *Global Climate Change Impacts in the U.S.* (2009). The scientific reports on climate change prepared by these organizations are the best, defendable science available to inform local climate action planning.

Based largely on these reports, this section summarizes the science of greenhouse gas emissions and climate change. This summary provides local government staff with strategies for communicating climate change, which will be needed to engage the various participants in the climate action planning process.

Planning typically begins with identifying and defining a problem that needs to be solved. For climate action planning the problem is global warming, which drives climate change (explained in the next section). Communities choose to address climate change for a variety of reasons, as explained in chapter 1, but once they commit to taking action, they must all be clear about how they define the problem. Good problem definition helps the public and decision makers understand the challenges presented by climate change and the role of a local response.

A Primer on Greenhouse Gas Emissions and Climate Change

The presence of gases in Earth's atmosphere, such as carbon dioxide (CO_2) and water vapor, create a natural greenhouse effect that traps heat generated by the sun. The greenhouse effect maintains a warm enough temperature for life to survive. Scientists have observed that Earth's climate has been warming since the late 1800s, most rapidly in the last few decades. The observations include an "increase in global average air and oceans temperatures, widespread melting of snow and ice, and rising global average sea level."[4] This warming is primarily attributed to an increase in greenhouse gases in the atmosphere caused by the burning of fossil fuels.[5] The rate of this warming is unprecedented and threatens to warm the planet by as much as three to seven degrees Fahrenheit by 2100. The potential impacts of this level of warming are numerous but most notably include the rise of sea level threatening low-lying coastal communities; an increase in the occurrence of extreme weather-related events such as heat waves, storms, floods, droughts, and wildfires; and the loss of native plants and wildlife. These impacts potentially threaten the social, economic, and cultural stability of many of the planet's communities.

How Do We Know the Earth Is Warming?

Extensive datasets gathered by multiple scientific organizations track a range of climate indicators such as temperature. Together, these data provide compelling evidence of global warming. The National Oceanic and Atmospheric Administration (NOAA) report *State of the Climate in 2009* identifies ten indicators of global warming based on these datasets.[6] All of the following indicators show increasing trends consistent with global warming:

- Land surface air temperature (based on four datasets)
- Tropospheric temperatures (based on seven datasets)
- Sea-surface temperature (based on six datasets)
- Marine air temperature (based on five datasets)
- Ocean heat content (based on seven datasets)
- Sea level (based on six datasets)
- Specific humidity (based on three datasets)

Decreasing trends in the following indicators are also consistent with global warming:

- Northern hemisphere snow cover (based on two datasets)
- Glacier mass balance (based on four datasets)
- September arctic sea-ice extent (based on three datasets)

The NOAA researchers conclude the following: "The observed changes in a broad range of indicators provide a self-consistent story of a warming world."[7]

Each of the measures represents a global average. In any specific area of the globe at any specific time these trends may be different. For example, in 2009 Europe experienced warmer than usual average temperatures, whereas the American Midwest experienced cooler than average temperatures. In addition, year to year the average global surface temperature varies up and down, but the long-term trend is upward. These year-to-year changes are due to annual variability in a variety of climate-related systems, particularly ocean currents, that affects global temperatures. This raises the issue of "weather" versus "climate." *Weather* typically refers to changes over a shorter period of time, and *climate* refers to changes over a longer period of time. A particularly cold week, month, or year is not evidence against global warming nor is a particularly hot week, month, or year evidence supporting global warming. The longer-term indicators established by NOAA (listed earlier) constitute the best evidence to date of global warming.

What Causes Global Warming?

The greenhouse effect is a natural phenomenon that has caused the temperature of the planet to be warmer than it otherwise would if it had no atmosphere. If the natural greenhouse effect did not occur, the average temperature of the planet would be about 60°F colder than it is currently. The natural greenhouse effect is a process wherein solar energy heats the planet, and some of this heat is radiated back toward space only to be absorbed by atmospheric gases and re-radiated back toward the planet's surface. In other words, the planet is essentially heated by two phenomena: the direct heating of the surface and atmosphere from the sun, and the "captured" heat that would have radiated back out to space by the atmosphere. Although the term *greenhouse effect* is used, the phenomenon is not exactly like a greenhouse, but the analogy works well enough. Earth can be thought of as a giant greenhouse, with an atmosphere that contains certain gases that act as the glass.

The naturally occurring greenhouse gases are carbon dioxide (CO_2), water vapor (H_2O), methane (CH_4), nitrous oxide (N_2O), and ozone (O_3). There are also several human–made chemicals that are released into the atmosphere and that act as greenhouse gases. These are generically called halocarbons (box A.1). The addition of these human-made chemicals and the addition of the naturally occurring greenhouse gases beyond their natural levels have contributed to anthropogenic (human-caused) global warming. In other words, humans have enhanced the natural greenhouse effect that made the planet a livable temperature. We have created thicker layers of glass in our greenhouse. As noted earlier, this additional warming could have a serious impact on the planet.

Each of these greenhouse gases does not have the same effect on the energy balance (or heat transfer) of the atmosphere (called radiative forcing). Some of these greenhouse gases have more global warming potential (GWP) and are thus of more concern than others. This is because of both their physical properties, their quantities in the atmosphere, and their longevity in the atmosphere. Carbon dioxide (CO_2) in the atmosphere has the highest relative radiative forcing and is considered a long-lived greenhouse gas, meaning that the carbon dioxide being put into the atmosphere will remain there for a long time. There are also several factors working to reduce anthropogenic global warm-

Greenhouse Gases

Carbon Dioxide (CO_2)

A naturally occurring gas that is also a by-product of burning fossil fuels and bio-mass, as well as land-use changes and other industrial processes. It is the principal anthropogenic (human-made) greenhouse gas that affects Earth's radiative balance between incoming and outgoing heat. It is the reference gas against which other greenhouse gases are measured and therefore has a global warming potential of 1.

Fluorocarbons

Carbon-fluorine compounds that often contain other elements such as hydrogen, chlorine, or bromine. Common fluorocarbons include chlorofluorocarbons (CFCs), hydrochlorofluorocarbons (HCFCs), hydrofluorocarbons (HFCs), and perfluorocarbons (PFCs). These compounds can be found in a variety of materials and processes such as air-conditioning, building materials, and industrial operations.

Halocarbons

Compounds containing either chlorine, bromine or fluorine, and carbon. Such compounds can act as powerful greenhouse gases in the atmosphere. The chlorine- and bromine-containing halocarbons are also involved in the depletion of the ozone layer. While there are natural sources of halocarbons, the addition of anthropogenic sources have led to warming. Halocarbons have many uses such as solvents, pesticides, refrigerants, adhesives, and specialized industrial uses.

Methane (CH_4)

A hydrocarbon with a global warming potential most recently estimated at 23 (although a previous estimate of 21 is still in common usage) times that of carbon dioxide (CO_2). Methane is produced through anaerobic (without oxygen) decomposition of waste in landfills, animal digestion, decomposition of animal wastes, production and distribution of natural gas and petroleum, coal production, and incomplete fossil fuel combustion.

Nitrous Oxide (N_2O)

A powerful greenhouse gas with a global warming potential nearly 300 times that of carbon dioxide (CO_2). Major sources of nitrous oxide include soil cultivation practices, especially the use of commercial and organic fertilizers, fossil fuel combustion, nitric acid production, and biomass burning.

Ozone (O$_3$)

Ozone (O$_3$) is a gaseous atmospheric constituent. Its contribution to global warming varies depending on the atmospheric layer in which it is located. In the troposphere, it is created both naturally and by photochemical reactions (pollutants + sunlight) involving gases resulting from human activities (smog). Tropospheric ozone acts as a greenhouse gas with higher concentrations resulting in warming. In the stratosphere, ozone is created by the interaction between solar ultraviolet radiation and molecular oxygen (O$_2$). Stratospheric ozone plays a decisive role in the stratospheric radiative balance. Depletion of stratospheric ozone allows increased solar radiation to reach Earth. In other words, ozone in the stratosphere reduces warming, and ozone in the troposphere increases it.

Water Vapor (H$_2$O)

The most abundant greenhouse gas, it is the water present in the atmosphere in gaseous form. Water vapor is an important part of the natural greenhouse effect. While humans are not significantly increasing its concentration, it contributes to the enhanced greenhouse effect because the warming influence of greenhouse gases leads to increased levels of water vapor. In addition to its role as a natural greenhouse gas, water vapor plays an important role in regulating the temperature of the planet because clouds form when excess water vapor in the atmosphere condenses to form ice and water droplets and precipitation.

Source: Compiled from the U.S. EPA glossary, http://www.epa.gov/climatechange/glossary.html.

ing. These include surface albedo (the ability of Earth's surface to reflect sunlight back into space before it causes warming) and aerosols, especially clouds (that also can reflect sunlight). Human activities have increased Earth's surface albedo, especially where forests (darker areas) have been converted to other uses (usually lighter) and have caused an increase in cloud development.

What Is the Difference between Global Warming and Climate Change?

According Robert Henson, author of *The Rough Guide to Climate Change*, the terms *global warming* and *climate change* have essentially become interchangeable in contemporary usage.[8] In general, *climate change* has become the preferred term partly to avoid the confusion that occurs with people who assume global warming means that every place will be warmer all the time. Climate change could be considered more accurate since the phenomenon will result in greater temperature variability and volatility, not just uniform warming in all parts of the globe. This book uses both terms but will most often prefer *climate change*.

What Are the Levels and Sources of Greenhouse Gases?

To develop a common metric for all greenhouse gases, scientists often refer to emissions in CO_2 equivalent units (abbreviated CO_2e, CO_2eq, eCO_2, or CDE) based on their GWP. Only 85% of the greenhouse gas emissions reported for the United States in 2008 were specifically CO_2, whereas the remainder were primarily methane and nitrous oxide.[9] Researchers convert the non-CO_2 gases into CO_2 equivalents (CO_2e) for the convenience of reporting and understanding the overall effect of greenhouse gas emissions. It is important to note that the GWP reported for various greenhouse gases varies depending on the source. GWP cannot be known precisely, which explains the variation seen in the climate science literature. While there is variation, the relative scale tends to be consistent. For example, methane is always reported as having a GWP between 20 and 25, and nitrous oxides are always in the vicinity of 300.

Atmospheric measures of CO_2 and CO_2e are usually made in parts per million volume (ppmv or ppm). As of 2010 atmospheric CO_2 was 390 ppm as measured by NOAA at the Mauna Loa Observatory, Hawaii (this roughly translates to 450 ppm CO_2e). This is up from 317 ppm in 1960 as measured at the same station. In addition, the IPCC states the following: "Current concentrations of atmospheric CO_2 and CH_4 far exceed pre-industrial values found in polar ice core records of atmospheric composition dating back 650,000 years."[10] A recent report contains estimates of the increase in atmospheric concentrations of greenhouse gases' effect on average global warming level: 340 ppm CO_2e roughly equates to a 1°C global average warming; 430 ppm CO_2e to 2°C; 540 ppm CO_2e to 3°C; 670 ppm CO_2e to 4°C; and, 840 ppm CO_2e to 5°C.[11]

Table A.1   Top 10 "key categories" of greenhouse gas emissions

Key category	% of total (2008 CO₂e)
CO_2 emissions from stationary combustion—coal	30.7
Mobile combustion: road and other	23.5
CO_2 emissions from stationary combustion—gas	17.6
CO_2 emissions from stationary combustion—oil	7.7
Direct N_2O emissions from agricultural soil management	2.5
Mobile combustion: aviation	2.3
CH_4 emissions from enteric fermentation	2.1
CO_2 emissions from non-energy use of fuels	2.0
CH_4 emissions from landfills	1.9
Emissions from substitutes for ozone depleting substances	1.7
Total of all other key categories	8.0

The primary source, but not the only source, of anthropogenic greenhouse gases is the burning of fossil fuels for energy and transportation. U.S. greenhouse gas emissions in 2008 were 6,957 Tg (teragrams) CO_2e (an increase of 14% since 1990), most of which came from the combustion of coal for electricity production (see fig. 1.1 in chap. 1).[12] The burning of petroleum for transportation is a close second in total emissions. As a result, these two sectors are usually the focus of climate action planning; they represent the largest part of the global warming problem. Table A.1 represents this information a little differently by using the "key categories" used by the U.S. Environmental Protection Agency (EPA) in the national inventory. In plain English, these key categories translate to the activities occurring every day in communities across the United States. People drive their cars and trucks, fly in airplanes, turn on electrical and gas appliances in their homes, especially for heating and cooling, eat food grown on farms, and throw away their garbage. Businesses, industries, farms, and government agencies do similar things and provide people with the goods and services they demand. All of these activities result in the production of greenhouse gases.

In addition to the direct emission of greenhouse gases, human activities can contribute to global warming by adversely impacting carbon sinks. A sink is anything that absorbs more carbon than it releases. Oceans, soils, and forests all store carbon and can act as sinks. The degree

Table A.2   Common greenhouse gases atmospheric lifetimes and global warming potential

Greenhouse gas	Average lifetime in the atmosphere	Global warming potential of one molecule of the gas over 100 years (relative to carbon dioxide = 1)
Carbon dioxide	50–200 years★	1
Methane	12 years	21
Nitrous oxide	120 years	310
CFC-12	100 years	10,600
CFC-11	45 years	4,600
HFC-134a	14.6 years	1,300
Sulfur hexafluoride	3,200 years	23,900

★ Carbon dioxide's lifetime is poorly defined because the gas is not destroyed over time, but instead moves between different parts of the ocean–atmosphere–land system. Some of the excess carbon dioxide will be absorbed quickly (e.g., by the ocean surface), but some will remain in the atmosphere for thousands of years.

Source: U.S. Environmental Protection Agency, *Climate Change Indicators in the United States*, EPA 430-R-10-007 (Washington, DC: Author, 2010), 18.

to which they serve as a sink varies seasonally within the global carbon cycle. Changes in climate and human development practices can influence the degree to which carbon can be stored. For example, increasing water temperature in oceans reduces the carbon storage capacity. A reduction in carbon uptake rate of the ocean is equivalent to direct emissions of CO_2 into the atmosphere. Similarly, deforestation and soil degradation both reduce the CO_2 uptake and storage capacity of ecosystems and result in increased atmospheric greenhouse gas concentrations. For the GWP of common greenhouse gases see table A.2.

What Are the Projected Climate Impacts?

Climate change directly alters temperature, sea level, and precipitation patterns. These changes result in a much wider range of consequences, including drought, wildfire, flooding, and species migration disruption. Each of these events has the potential to disrupt or damage local resources and systems, including the built environment (infrastructure and buildings), economic and social resources, and ecosystems. Climate change impacts can be expressed on large spatial scales as alteration in the probability and magnitude of event occurrence (e.g., 100-year flood or extended heat wave). The impacts experienced by communities will vary widely. This is a challenge for local jurisdictions, which must communicate the likely local consequences of climate change and formulate policy to reduce local vulnerability. Brief summaries of the climate change impacts projected for the United States and some of the consequences of these changes follow.[13]

Sea Level Rise

Two processes drive sea level rise: the melting of glaciers and thermal expansion of marine waters. Sea level has risen 8 inches in the last 100 years and is projected to continue at nearly double the historic rate. Melting of Antarctic or Greenland ice sheets may result in much more extensive impacts, but the likelihood and timing of these events are uncertain. Sea level rise poses a series of consequences such as coastal flooding (gradual coastal inundation and coastal stream flooding), extreme high tide, increased erosion, ecosystem loss (estuaries), and saltwater intrusion into groundwater.

These changes can have a range of impacts on community resources along the coast. Many critical components of a coastal community's infrastructure can be located near the coast. Transportation infrastructure such as roads, marine ports, airports, and train lines located in low-lying coastal areas are potentially vulnerable to sea level rise impacts. In addition to transportation networks, population often concentrates near coastal areas. This concentration of structures results in sea level rise threatening not only buildings but also the safety of inhabitants. Sea level rise can also pose a threat to local water supplies through intrusion into groundwater resources and damage to local ecosystems. These two impacts may have economic consequences from the viability of drinking water sources to losses due to recreation.

Temperature Variation

Changes in the temperature can mean different outcomes depending on location. In some locations this may mean a higher frequency of high-temperature days (> 90° or 100°F) and prolonged periods of extreme heat (heat waves). In others it may not result in extreme heat but will alter the timing and duration of seasons such as shorter winters with fewer cold days. These changes have the potential to result in a wide range of community outcomes. Heat has direct consequences for human health from lethargy to heat stroke to even death. Heat can result in increased formation of ground-level ozone, which is associated with many respiratory ailments. There are also interactions between heat outcomes. Heat may result in increased water and energy use. It may also increase fire frequency and severity. Fire can pose a direct threat to human safety and structures as well as reduce air quality and threaten ecosystems.

Changes in annual temperature patterns can result in changes in natural systems and processes. Chemical reactions such as those in water change with temperature, meaning water quality could shift. In addition, changing seasons will alter growing seasons, which has consequences for overall ecosystem function and agricultural production. Altered season length and average temperature can also result in increased pest and/or invasive species presence.

Precipitation Variation

Similar to the other climate impacts, change in annual amount and timing of precipitation varies based on location. The total amount of precipitation may change, as well as the timing, intensity, and form (e.g., snow, rain, etc.). These changes can result in reductions in rainfall or extended periods without rain (e.g., drought). Conversely, climate change can result in intensive storms that can cause flooding and erosion. The change in precipitation type can result in a reduction in snow and an increase in rain totals.

These changes, particularly when paired with alteration in temperature, can result in water scarcity for drinking and agricultural uses. The intensive rain can damage flood-prone areas, threatening structures, inhabitants, and infrastructure in low-lying areas. Alteration in precipitation can also impact ecosystem function from changing the availability of water to support plant life in stream ecosystems. Changing amount and timing of rivers will impact not only high flows but also low flows. Reduced flow conditions can make habitat unsuitable for aquatic species such as anadromous fish.

Glossary of Terms[14]

The following list includes some of the key terms often found in climate science articles and reports. Planners faced with communicating with the public and decision makers about the challenge presented by climate change should be familiar with these terms.

Albedo The fraction of incoming solar radiation reflected by a surface or object, often expressed as a percentage. Snow-covered surfaces have a high albedo, indicating a high level of reflectivity; the albedo of soils ranges from high to low; vegetation-covered surfaces and oceans have a low albedo. Earth's albedo is influenced by varying levels of cloudiness, snow, ice, leaf area, and land cover changes.

Anthropogenic Made by people or resulting from human activities. Usually used in the context of emissions that are produced as a result of human activities.

Atmosphere The gaseous envelope surrounding Earth. The dry atmosphere consists almost entirely of nitrogen (78.1% volume mixing ratio) and oxygen (20.9% volume mixing ratio), together with a number of trace gases, such as argon (0.93% volume mixing ratio), helium, and radiatively active greenhouse gases. Radiatively active gases such as carbon dioxide (0.035% volume mixing ratio), and ozone influence how much energy leaves the atmosphere. In addition the atmosphere contains water vapor, the amount of which is highly variable but is typically 1% volume mixing ratio.

Carbon dioxide equivalent (CO_2e, CO_2eq, eCO_2, and CDE) A measure used to compare the emissions from various greenhouse gases based upon their global warming potential (GWP). Carbon dioxide equivalents are commonly expressed as million metric tons of carbon dioxide equivalents (MMTCDE) or million short tons of carbon dioxide equivalents (MSTCDE). The carbon dioxide equivalent for a gas is derived by multiplying the tons of the gas by the associated GWP. MMTCDE = (million metric tons of a gas) × (GWP of the gas). For example, the GWP for methane is 21. This means that emissions of 1 million metric tons of methane is equivalent to emissions of 21 million metric tons of carbon dioxide.

Climate Climate, in a narrow sense, is usually defined as the "average weather." A more rigorous statistical description is the mean and variability of quantities such as temperature, precipitation, and wind over a period of time ranging from months to thousands of years. The most commonly used period is three decades.

Climate change Climate change refers to any significant change in measures of climate (such as temperature, precipitation, or wind) lasting for an extended period (decades or longer). Climate change may result from natural factors (e.g., changes in the sun's intensity or slow changes in Earth's orbit around the sun), natural processes within the climate system (e.g., changes in ocean circulation), and human activities that change the atmosphere's composition (e.g., through burning fossil fuels) and the land surface (e.g., deforestation, reforestation, urbanization, desertification, etc.).

Global warming Global warming is an average increase in the temperature of the atmosphere near Earth's surface and in the troposphere, which can contribute to changes in global climate patterns. Global warming can occur from a variety of causes, both natural and human induced. In common usage, *global warming* often refers to the warming that can occur as a result of increased emissions of greenhouse gases from human activities.

Global warming potential (GWP) The cumulative radiative forcing effects from the emission of a unit mass of gas relative to a reference gas. The GWP-weighted emissions of direct greenhouse gases in the U.S. Inventory are presented in terms of equivalent emissions of carbon dioxide (CO_2).

Greenhouse effect The greenhouse effect is the trapping and buildup of heat in the atmosphere (troposphere) near Earth's surface. Some of the heat flowing back toward space from Earth's surface is absorbed by water vapor, carbon dioxide, ozone, and several other gases in the atmosphere and then radiated back toward Earth's surface. If the atmospheric concentrations of these greenhouse gases rise, the average temperature of the lower atmosphere will gradually increase. Greenhouse gases include, but are not limited to, water vapor, carbon dioxide (CO_2), methane (CH_4), nitrous oxide (N_2O), chlorofluorocarbons (CFCs), hydrochlorofluorocarbons (HCFCs), ozone (O_3), hydrofluorocarbons (HFCs), perfluorocarbons (PFCs), and sulfur hexafluoride (SF_6).

Enhanced greenhouse effect The anthropogenic emissions of greenhouse gases have served to enhance or amplify the natural greenhouse effect. Increased concentrations of carbon dioxide, methane, and nitrous oxide, chlorofluorocarbons (CFCs), hydrochlorofluorocarbons

(HCFCs), perfluorocarbons (PFCs), sulfur hexafluoride (SF_6), nitrogen trifluoride (NF_3), and other heat-trapping gases caused by human activities such as fossil fuel consumption, trap more infrared radiation, thereby exerting a warming influence on the climate.

Radiative forcing Radiative forcing refers to actions that impact the energy balance of the planet. In equilibrium, the same amount of energy that enters the system (sunlight) would leave (emitted as heat). If there is any imbalance in the energy entering and leaving the atmosphere, the earth would be heating or cooling. The presence or absence of this balance is most meaningfully measured at the boundary between the troposphere (the lowest level of the atmosphere) and the stratosphere (the thin upper layer). Radiative forcing is a way to measure the impact of human activities on global temperature. It is influenced by the level of greenhouse gases present, as well as changes in albedo (surface reflectivity), clouds, and solar input. Additional detail on radiative forcing and its measurement can be found in chapter 6 of the 2001 IPCC *Third Assessment Report Working Group I: The Scientific Basis*.[15]

Troposphere The lowest part of the atmosphere from Earth's surface to about 10 km in altitude in midlatitudes (ranging from 9 km in high latitudes to 16 km in the tropics, on average) where clouds and weather phenomena occur. In the troposphere temperatures generally decrease with height.

Weather Atmospheric condition at any given time or place. It is measured in terms of such things as wind, temperature, humidity, atmospheric pressure, cloudiness, and precipitation. In most places, weather can change from hour to hour, day to day, and season to season. Climate can be defined as the "average weather." A simple way of remembering the difference is that climate is what you expect (e.g., cold winters) and weather is what you get (e.g., a blizzard).

Resources

Books

Robert Henson, *The Rough Guide to Climate Change*, 2nd ed. (New York, NY: Rough Guides, 2008). This book provides an introduction to the science of climate change that is appropriate for citizens and public officials. In addition to the science, it includes a section on what individuals can do to reduce their carbon footprint.

Organizations

Intergovernmental Panel on Climate Change (IPCC). http://www.ipcc.ch/. The IPCC website provides the organization's various assessment reports covering physical science; impacts, adaptation, and vulnerability; and mitigation. The most recent reports are the 2007 *Fourth Assessment Report* (AR4) series. The most accessible of these is the *Synthesis Report Summary for Policymakers.* http://www.ipcc.ch/publications_and_data /ar4/syr/en/spm.html.

U.S. Global Change Research Program (USGCRP). http://www .globalchange.gov/. The USGCRP website contains numerous reports on the science of climate change including an annual report titled *Our Changing Climate* and the 2009 *Global Climate Change Impacts in the U.S.,* the latter being a key report for understanding the expected impacts of climate change. In addition the website describes the roles of thirteen federal agencies and departments in addressing climate change.

U.S. Environmental Protection Agency (EPA). http://www.epa.gov /climatechange/. The EPA website contains information on climate change indicators, science, greenhouse gas emissions, health and environment, economics, regulatory initiatives, and public policy.

Examples of Science Presentation in Climate Action Plans

City of Homer, Climate Action Plan (2007). http://www.ci.homer.ak.us /CLPL.pdf. This plan is an example of how to present climate change science in a way that is relevant to the local community. It also establishes the influence of the science in motivating the mayor and city council to act.

City of Chicago, Climate Action Plan (n.d.). http://www.chicagoclimate action.org/. This plan is an example of a concise, graphics-driven approach to presenting key ideas on climate change science without overwhelming the reader with too much detail.

City of Cambridge, Climate Protection Plan (n.d.). http://www .cambridgema.gov/cdd/et/climate/. This plan is an example of how to explain the science and impacts of climate change in a way that argues for action. The plan contains sections on "Why Waiting Is Not an Option" and "Reasons to Take Action."

Appendix B:
The Public Participation Program

This section provides an outline for a model public participation program that can be tailored based on the public participation approach desired and the answers to the key questions presented in chapter 3. Similar to the climate action planning process, a public participation program can be seen as including three phases with a number of actions in each phase:

1. Preliminary phase
   a. Establish goals for the program.
   b. Develop a target audience list and identify stakeholders.
   c. Create key messages and an "identity."
   d. Publicize the climate action planning process (media, website, social media, e-mail, events).
2. Planning phase
   a. Kickoff event
   b. Communications
   c. Workshops/task forces meetings/focus groups
3. Adoption and implementation phase
   a. Adoption meetings
   b. Celebration
   c. Implementation activities

Preliminary Phase

The preliminary phase includes the tasks that should be completed before engaging the public in the planning process.

Goals for the Program

The goals of the public participation program for the climate action plan (CAP) include the following:

- Communicate to the community about the purpose of the CAP and the impacts of CAP implementation on the three primary sectors (energy, transportation, waste) and provide opportunities for the community to provide input as to how emissions reduction goals should be reached.
- Promote the CAP according to the message or issue of importance to the community, rather than the project.
- Position the planning organization as the best resource for information about the CAP.
- Generate interest and identify early supporters.
- Ensure community empowerment, buy-in, and long-term success.

Target Audience

While the whole community is ultimately the outreach target, sub-populations of the whole are concerned about different issues and will require different techniques of engagement. The following are some of the key populations:

- Local business owners
- Environmental advocacy groups
- Homeowners
- Retirees
- Residents who rent their home
- Utilities
- Partner agencies
- Community-based organizations
- Ethnic and cultural groups
- Development stakeholders

Key Messages and an "Identity"

Key messages are the main points to convey to stakeholders, from residents to policymakers. These messages are crafted to move the community to action, in this case to attend public meetings and provide input on transportation, energy, waste, land use, and water policies. How these issues relate to the community's response to the potential impacts of climate change is of paramount importance. Key messages should remain simple and straightforward. To achieve this they can also be divided into primary and secondary messages if necessary.

Important to the completion of the CAP will be communicating these messages to the local and regional community. One effective way to do this is by creating an identity or "look-and-feel" through design choices, slogans, and iconic images, and a corresponding outreach program that promotes the CAP. There is an important distinction to be made between the development of an identity and key messages. An identity typically includes a graphic representation that conveys something permanent about the CAP. Unlike key messages, which are nimble and vary depending on the issue and the target audience, an identity would be set in place to convey one message to every target audience for a lifetime of at least five years.

Publicity

The CAP process should be publicized in a diverse manner in order to reach the maximum number of community members. This process should begin with identification of local social and cultural hubs, local events, and familiar communication networks. Press releases, announcements, and supporting materials should be developed to reflect non-English-speaking populations in the community. Announcements should be made using the radio, the Internet, posters at community gathering points, and displays at local events.

A website or webpage should be developed after identification of the CAP's identity and messaging. The decision to create a stand-alone project site or a webpage integrated in the agency's existing website is based on the agency's existing online tools and resources. If the agency has a known presence online, such as the provision of green or sustainability resource webpages, then the CAP should be integrated into existing resources. If the agency is using the CAP to launch a larger sustainability effort, then a stand-alone website would be beneficial.

Planning Phase

In the planning phase the public is engaged through several measures to educate them on climate change and solicit their input on the CAP.

Kickoff Event (Workshop One)

The CAP should be launched at a high profile and interactive event. Depending on the community, the launch or kickoff event could include a town hall style meeting or workshop, a mobile workshop or booth at an established event (i.e., weekly farmer's market or other regular community event), or as an open house. The event should be heavily promoted; media outreach as well as public outreach will be necessary to reach a broad spectrum of stakeholders. The goal of the kickoff event is to inform community stakeholders about the CAP and planning process. Objectives should be to provide a meeting approach that balances education, engagement, and input. Tools may include a mix of traditional large group presentations, nontraditional polling, and small-group exercises. The kickoff event should occur early in the project, often in the first two months. The event should provide an overview of the CAP—the planning process and project objectives—as well as any relevant local background information, such as the results of the community-wide greenhouse gas (GHG) inventory or potential impacts/risks of climate change. The event should include opportunities to present information (e.g., educate or inform) and to receive input. Information should be presented with visual aids, including Power-Point presentations, boards with graphics, or videos. Opportunities for participants to provide input or engage include real-time polling, question and answer sessions, small-group facilitated tables, or information stations. A computer-based polling system (e.g., Turning Point) may be used to gather anonymous and immediate feedback on existing conditions and future policy direction.

Communications

Organize a speakers bureau following the kickoff event and in advance of the additional workshops that will enable plan managers to connect with key stakeholder groups on important issues. The speakers bureau will include development of a PowerPoint presentation on the CAP sectors and process. Speakers should be selected carefully as they will be the ambassadors of the project. The plan proponent should provide training to all speakers and opportunities for feedback and support.

Develop a stakeholder database and send a minimum of two blast e-mails (e-blasts) in advance of the kickoff event and all workshops to promote community attendance. Developing the list to include a cross-section of stakeholders in the community will be important. The stakeholder list will also require regular updating as people sign up via the website, Facebook, or at public meetings. Also, send e-blasts to announce website updates, release of drafts, and public hearings. Stakeholder and media outreach begins four to six weeks in advance of the kickoff event (and all workshops).

Develop an online or telephone survey to accompany the initial planning process. The survey will alert participants and stakeholders to the CAP planning process and also provide an opportunity for stakeholders to offer their level of education about the CAP, climate change, and the contribution of their individual behaviors to GHGs. The survey can also assess willingness to change and priorities for GHG reduction strategies. The online survey should use a standardized online survey software program. Online surveys are not usually statistically valid; however, they do offer information useful to the planning process. Telephone surveys provide statistically valid results and could be more applicable for high profile or sensitive topics. The survey should activate concurrent with the public outreach for Workshop One. The survey should end prior to development of draft GHG reductions strategies, at least one month prior to Workshop Two.

Workshops/Task Force Meetings/Focus Groups

Workshop Two, Focus on GHG Reduction Measures, will be a more traditional public workshop in which participants engage in a variety of activities where they can provide input and feedback on the GHG reduction measures. Opportunities for facilitated small group discussions will be provided for examination of specific issues. For the second workshop, provide up to six large-format posters for use during small group discussions and handouts that summarize the reduction measures by sector.

In addition to the main workshops, stakeholder roundtables or mini-workshops provide an opportunity for key stakeholders to provide input on specific sectors in a facilitated setting. The meetings are suitable for up to twenty participants and most productive when stake-holders are separated by the sectors of the GHG inventory and CAP—energy, land use, transportation, agriculture, business. For each meeting, participants should receive an information packet including background information and preliminary GHG reduction strategies (or best practices), questions for discussion, and ground rules for participation. The meeting format should include a presentation and facilitated brain-storming and discussion. The objectives and next steps should be clear to all participants.

Dispute resolution is needed in cases of disagreement or conflict over an issue. Bring in professional mediators to help the parties resolve their dispute.

For facilitated brainstorming and consensus-building, bring in professional facilitators to help the citizens brainstorm ideas and develop agreements about the best ideas.

Workshop Three, Reviewing the Draft CAP, will follow a format similar to Workshop Two. The objective of the workshop will be to provide an opportunity for the public to provide input and feedback on the draft CAP before the public hearing process. Opportunities for facilitated small group discussions will be provided for discussion of specific issues.

Adoption and Implementation Phase

The adoption and implementation phase includes the tasks to be completed once a draft of the plan is ready for review and adoption by decision-makers.

Adoption Meetings

Schedule at least two open, noticed public meetings or hearings before the adopting government board or entity. In some cases, a joint study session of the planning commission and city council provides an opportunity for questions and discussions prior to an adoption hearing. Prepare presentation and handouts summarizing the CAP. Provide electronic copies of the CAP online. Ensure that the draft CAP is available at least one month prior to the meetings.

Celebration

After adoption, create a community event to celebrate the CAP, inform the public about actions they can take, and kick off key implementation strategies. The event should be scheduled for a Saturday in a public area and include activities for all ages.

Implementation Activities

Continue to involve the public in decisions about implementation. Develop an annual reporting or review process that informs the public about progress and obtains feedback. Maintaining participation of key community stakeholders throughout implementation will require a mix of online and traditional information sharing, education, and interactive tools. Online tools allow the agency to report on implementation progress and allow for individual tracking and contribution to the overall target.

Develop or support community-based or peer-to-peer education and networking forums to facilitate implementation of reduction measures that rely on changes in business-as-usual practices.

Notes

Chapter 1

1. "1,000th Mayor—Mesa, AZ Mayor Scott Smith Signs the U.S. Conference of Mayors Climate Protection Agreement," The U.S. Conference of Mayors, October 2, 2009, accessed August 23, 2010, http://www.usmayors.org/pressreleases/uploads/1000signatory.pdf.

2. Thomas R. Karl, Jerry M. Melillo, and Thomas C. Peterson, eds., *Global Climate Change Impacts in the United States* (New York: Cambridge University Press, 2009).

3. "The U.S. Mayors Climate Protection Agreement," The U.S. Conference of Mayors, accessed February 19, 2011, http://www.usmayors.org/climateprotection/revised/.

4. "1,000th Mayor—Mesa, AZ Mayor Scott Smith Signs the U.S. Conference of Mayors Climate Protection Agreement," The U.S. Conference of Mayors, October 2, 2009, accessed August 23, 2010, http://www.usmayors.org/pressreleases/uploads/1000signatory.pdf.

5. "Mayor Davlin Signs Climate Protection Agreement: Springfield Becomes Cool City," City of Springfield, Illinois, August 26, 2008, accessed August 23, 2010, http://www.springfield.il.us/releases/2008%20Releases/CoolCities.htm.

6. Count based on an inventory by the authors of stand-alone climate action plans of ICLEI and non-ICLEI member communities. . . . ICLEI reports similar numbers, although they include sustainability plans.

7. In 2003 the organization changed its name from the International Council for Local Environmental Initiatives to "ICLEI–Local Governments for Sustainability," or ICLEI for short.

8. ICLEI-USA, *Measuring Up: A Detailed Look at the Impressive Goals and Climate Action Progress of U.S. Cities & Counties,* 2009 Annual Report (Boston, MA: Author, 2009).

9. American College & University Presidents' Climate Commitment, accessed February 19, 2011, http://www.presidentsclimatecommitment.org/.

10. Compiled from: Natural Capital Solutions, *Climate Protection Manual for Cities* (Eldorado Springs, CO: Author, 2007). American Planning Association, *Policy Guide on Planning and Climate Change* (Washington, DC: Author, 2008). National Wildlife Federation, *Guide to Climate Action Planning:*

Pathways to a Low-Carbon Campus (Reston, VA: Author, 2008). International Council for Local Environmental Initiatives, *U.S. Mayors' Climate Protection Agreement: Climate Action Handbook,* accessed May 29, 2009, http://www.seattle.gov/climate/docs/ClimateActionHandbook.pdf. International Council for Local Environmental Initiatives, *ICLEI Climate Program,* accessed May 11, 2010, http://www.iclei.org/index.php?id=800.

11. GHG emissions are measured in metric tons of carbon dioxide (CO_2) equivalents, designated CO_2e or eCO_2. Since there are several greenhouse gases in addition to CO_2, a common metric has been developed to put them in equivalent units based on their relative levels of contribution to global warming. For example, 1 unit of methane is equivalent to 25 units of carbon dioxide.

12. U.S. Environmental Protection Agency State and Local Climate and Energy Program, *A Quick Guide to GHG inventories,* EPA-430-F-09-003 (Washington, DC: EPA, 2009).

13. City of Cincinnati, *Climate Protection Action Plan: The Green Cincinnati Plan* (Cincinnati: Office of Environmental Quality, 2008).

14. Stephen M. Wheeler, "State and Municipal Climate Change Plans: The First Generation," *Journal of the American Planning Association* 74, no. 4 (2008): 481–96. American Planning Association, *Policy Guide on Planning and Climate Change* (Washington, DC: Author, 2008). International Council for Local Environmental Initiatives, *U.S. Mayors' Climate Protection Agreement: Climate Action Handbook,* accessed May 29, 2009, http://www.seattle.gov/climate/docs/ClimateActionHandbook.pdf.

15. Leslie Kaufman, "In Kansas, Climate Skeptics Embrace Cleaner Energy," *New York Times,* October 18, 2010, http://www.nytimes.com/2010/10/19/science/earth/19fossil.html?_r=2&hp.

16. "Organization," Intergovernmental Panel on Climate Change, accessed May 10, 2010, http://www.ipcc.ch/organization/organization.htm.

17. Intergovernmental Panel on Climate Change, "Human and Natural Drivers of Climate Change," *Climate Change 2007: Working Group I: The Physical Science Basis,* accessed June 1, 2011, http://www.ipcc.ch/publications_and_data/ar4/wg1/en/spmsspm-human-and.html.

18. Intergovernmental Panel on Climate Change, "Summary for Policymakers," in *Climate Change 2007: The Physical Science Basis. Contribution of Working Group I to the Fourth Assessment Report of the Intergovernmental Panel on Climate Change,* ed. S. Solomon, D. Qin, M. Manning, Z. Chen, M. Marquis, K. B. Averyt, M. Tignor and H. L. Miller (Cambridge and New York: Cambridge University Press, 2007), 5.

19. Ibid., 10.

20. Ibid., 12.

21. Ibid., 13.

22. Intergovernmental Panel on Climate Change, "Summary for Policymakers," in *Climate Change 2007: Impacts, Adaptation and Vulnerability. Contribution of Working Group II to the Fourth Assessment Report of the Intergovernmental Panel on Climate Change*, ed. M. L. Parry, O. F. Canziani, J. P. Palutikof, P. J. van der Linden, and C. E. Hanson (Cambridge: Cambridge University Press, 2007), 9.

23. Ibid., 17.

24. Thomas R. Karl, Jerry M. Melillo, and Thomas C. Peterson, eds., *Global Climate Change Impacts in the United States* (New York: Cambridge University Press, 2009).

25. "What Will Climate Change Mean to Alaska?" State of Alaska, accessed September 3, 2010, http://www.climatechange.alaska.gov/cc-ak .htm.

26. "The Impact of Climate Change on South Carolina," State of South Carolina, Department of Natural Resources, accessed September 3, 2010, http:// www. dnr. sc. gov/climate /sco/ Publications/climate_change _impacts.php.

27. Fredrich Kahrl and David Roland-Holst, *California Climate Risk and Response*, Research Paper No. 08102801 (Berkeley: University of California, Department of Agricultural and Resource Economics, 2008).

28. City of Miami, *MiPlan: City of Miami Climate Action Plan* (June 2008).

29. City of Aspen, *Canary Initiative: Climate Action Plan* (2007).

30. United Nations, *Kyoto Protocol to the United Nations Framework Convention on Climate Change*, 1998, accessed September 3, 2010, http://unfccc .int/resource/docs/convkp/kpeng.pdf.

31. United Nations Framework Convention on Climate Change, *National Greenhouse Gas Inventory Data for the Period 1990–2007*, 2009, accessed September 3, 2010, http://unfccc.int/resource/docs/2009/sbi/eng /12.pdf.

32. United Nations Framework Convention on Climate Change, *Copenhagen Accord,* 2009, accessed September 3, 2010. http://unfccc.int/.

33. United Nations Framework Convention on Climate Change, *Quantified Economy-Wide Emissions Targets for 2020*, 2010, accessed April 25, 2010, http://unfccc.int/home/items/5264.php.

34. "U.S. Climate Policy Maps," Pew Center on Global Climate Change, accessed February 19, 2011, http://www.pewclimate.org/what _s_being_done/in_the_states/state_action_maps.cfm.

35. Stephen M. Wheeler, "State and Municipal Climate Change Plans: The First Generation," *Journal of the American Planning Association* 74, no. 4 (2008): 481–96.

36. Peter Newman, Timothy Beatley, and Heather Boyer, *Resilient Cities: Responding to Peak Oil and Climate Change* (Washington, DC: Island Press, 2009).

37. Lawrence D. Frank, Sarah Kavage, and Bruce Appleyard, "The Urban Form and Climate Change Gamble," *Planning* 73, no. 8 (2007): 18–23.

38. Stephen Pacala and Robert Socolow, "Stabilization Wedges: Solving the Climate Problem for the Next 50 Years with Current Technologies," *Science* 305, no. 5686 (2004): 968–72.

39. James H. Swara, "Local Government Action to Promote Sustainability," Alliance for Innovation, February 1, 2011, accessed March 1, 2011, http://transformgov.org/en/Article/100947/Local_Government_Action_to _Promote_Sustainability.

40. California has several laws and legal precedents that have made climate action planning almost mandatory. This partially explains why California cities are the largest share of adopted CAPs.

41. Michelle Hoogendam Cash, *Houston Market Overview*, accessed February 11, 2011, http://www.law.northwestern.edu/career/jobsearch/ documents/Houston_Market_Report.pdf.

42. City of Houston, Emissions Reduction Plan (August 2008).

43. "Top 20 Local Governments," U.S. Environmental Protection Agency, accessed February 20, 2011, http://www.epa.gov/greenpower /toplists/top20localgov.htm.

44. U.S. Environmental Protection Agency, "EPA Issues Second Annual Ranking of U.S. Cities with the Most Energy Efficient Buildings," accessed February 20, 2011, http://yosemite.epa.gov/opa/admpress.nsf /f0d7b5b28db5b04985257359003f533b/b3f51c0c0396abc1852576ee0072db 8d!OpenDocument.

45. Wendy Siegle, "Houston Scores Federal Grant to Reduce Transportation-Related Emissions," KUHK.com, accessed February 20, 2011, http://app1.kuhf.org/houston_public_radio-news-display.php?articles_id =1284069400.

46. "Stamford Cool & Green 2020—Environmental Accomplishments," City of Stamford, accessed September 4, 2010, http://www.cityof stamford.org/content/25/50/105109/109156.aspx.

47. City of Key West, Climate Action Plan (October 2009), 3.

48. Ibid., 27.

49. City of Santa Cruz, Draft Climate Action Plan (September 2010), 13.

50. Ibid., 24.

51. Ibid., 15.

52. J. M. Brown, "Santa Cruz's Climate Action Plan up for Review Tuesday: Report Says Local Jobs, Solar Technology Would Cut Green-

house Gas Emissions by 30 Percent," *Mercury News*, September 3, 2010, accessed September 3, 2010, http://www.mercurynews.com/breaking-news/ci _15986551.

Chapter 2

1. "The Five Milestone Process," ICLEI, accessed February 20, 2011, http://www.iclei.org/index.php?id=810.
2. Michael R. Boswell, William J. Siembieda, and Kenneth C. Topping, *Local Hazard Mitigation Planning in California,* report prepared for the State of California Governor's Office of Emergency Services (2008).
3. City of Chicago, Building Healthy, Smart and Green—Chicago's Green Building Agenda (Chicago: Author, 2005).
4. City of Chicago, *Chicago Climate Action Plan—What Chicago Is Doing,* accessed August 9, 2010, http://www.chicagoclimateaction.org/pages /energy_efficient_buildings/43.php.
5. U.S. Environmental Protection Agency, *Inventory of U.S. Greenhouse Gas Emissions and Sinks: 1990–2008 Public Review Draft* (Washington, DC: Author, 2010).
6. City of Benicia, California, Climate Action Plan (2009).

Chapter 3

1. Pew Partnership for Civic Change, *Ready, Willing, and Able: Citizens Working for Change* (Charlottesville, VA: Author, 2000), accessed August 23, 2010, http://www.pew-partnership.org/timeline.html.
2. Public opinion statements based on the following surveys: Stanford University, *Global Warming Poll,* Conducted by Abt SRB1 (July 2010). The Pew Research Center for The People and The Press, *Modest Support for "Cap and Trade" Policy: Fewer Americans See Solid Evidence of Global Warming* (October 22, 2009). Yale Project on Climate Change Communication and George Mason University Center for Climate Change Communication, *Climate Change in the American Mind: Americans' Global Warming Beliefs and Attitudes in June 2010* (June 2010). Frank Newport, *Americans' Global Warming Concerns Continue to Drop* (March 11, 2010).
3. E. Maibach, C. Roser-Renouf, and A. Leiserowitz, *Global Warming's Six Americas 2009: An Audience Segmentation Analysis* (Yale Project on Climate Change and the George Mason University Center for Climate Change Communication, 2009), accessed August 31, 2010, http://www .americanprogress.org/issues/2009/05/pdf/6americas.pdf.

4. Roger Pielke Jr., *The Climate Fix: What Scientists and Politicians Won't Tell You about Global Warming* (Philadelphia, PA: Basic Books, 2010).

5. William R. L. Anderegga, James W. Prall, Jacob Harold, and Stephen H. Schneider, "Expert Credibility in Climate Change." Proceedings of the National Academy of Sciences of the United States of America (2010) accessed June 23, 2010, http://www.pnas.org/content/early/2010/06/04 /1003187107.full.pdf+html.

6. Peter T. Doran and Maggie Kendall Zimmerman, "Examining the Scientific Consensus on Climate Change," *EOS* 90, no. 3 (2009): 22–23, accessed June 23, 2010, doi:10.1029/2009EO030002, http://tigger.uic.edu /~pdoran/012009_Doran_final.pdf.

7. Matthew C. Nisbett and Dietram A. Scheufele, "What's Next for Science Communication? Promising Directions and Lingering Distractions," *American Journal of Botany* 96, no. 1 (2009): 1767–78.

8. Samuel D. Brody, David R. Godschalk, and Raymond J. Burby, "Mandating Citizen Participation in Plan Making: Six Strategic Planning Choices," *Journal of the American Planning Association* 69, no. 3 (2003): 245–64.

9. U.S. Department of Energy, *How to Design a Public Participation Program* (EM-22) (Washington, DC: n.d.).

10. "Spectrum of Public Participation," International Association for Public Participation, accessed August 23, 2010, http://www.iap2.org/display common.cfm?an=5.

11. Jeffrey M. Berry, Kent E. Portney, and Ken Thomson, *The Rebirth of Urban Democracy* (Washington, DC: Brookings Institution Press, 1993), 55.

12. Ibid., 55.

13. Sherry Arnstein, "A Ladder of Citizen Participation," *Journal of the American Institute of Planners* 35, no. 4 (1969): 216–24.

Chapter 4

1. In addition to community-level GHG inventories, there are national, state, regional, corporate, facility, project, and product (i.e., life cycle) GHG inventories. These types of inventories are not discussed in this chapter.

2. U.S. Environmental Protection Agency State and Local Climate and Energy Program. *From Inventory to Action: Putting Greenhouse Gas Inventories to Work*, EPA–430–F–09–002. (Washington, DC: Author, 2009).

3. California Air Resources Board, California Climate Action Registry, ICLEI–Local Governments for Sustainability, and The Climate Registry, *Local Government Operations Protocol for the Quantification and Reporting of Greenhouse Gas Emissions Inventories Version 1.1* (May 2010).

4. The authors anticipate that the Community-Scale Greenhouse Gas (GHG) Emissions Accounting and Reporting Protocol will become the standard protocol used to guide preparation of community-wide baseline inventories. The Protocol is being developed by ICLEI–Local Governments for Sustainability USA and a team of stakeholders. It is scheduled for completion in late 2011.

5. The California Climate Registry and the Climate Registry protocols are not discussed separately as their General Reporting Protocol for governments is the *Local Government Operations Protocol*.

6. The California Climate Action Registry is a program of the Climate Action Reserve and serves as a voluntary greenhouse gas (GHG) registry to protect and promote early actions to reduce GHG emissions by organizations. The California Registry stopped accepting inventory reports in 2010. California Registry members are in the process of transitioning to the Climate Registry, its sister organization, for reporting.

7. The Climate Registry is a nonprofit collaboration among North American states, provinces, territories, and Native Sovereign Nations that sets consistent and transparent standards to calculate, verify, and publicly report greenhouse gas emissions into a single registry.

8. City of Aspen, *Canary Initiative, Climate Action Plan 2007–2009* (2007).

9. California Air Resources Board, California Climate Action Registry, ICLEI–Local Governments for Sustainability, and The Climate Registry, *Local Government Operations Protocol for the Quantification and Reporting of Greenhouse Gas Emissions Inventories Version 1.1* (May 2010).

10. Portions of this section were modified from Michael R. Boswell, Adrienne I. Greve, and Tammy L. Seale, "An Assessment of the Link between Greenhouse Gas Emissions Inventories and Climate Action Plans," *Journal of the American Planning Association* 76, no. 4 (2010): 451–62.

11. L. H. Sun, "Public Transit Ridership Rises to Highest Level in 52 Years," *Washington Post* (March 9, 2009), accessed June 9, 2010, http://www.washingtonpost.com/wp-dyn/content/article/2009/03/08/AR2009030801960.html.

12. League of American Bicyclists, "43% Increase in Bicycle Commuting since 2000," accessed October 20, 2009, http://www.bikeleague.org/blog/2009/09/43 increase-in-bicycle-commuting-since-2000/.

13. "U.S. Grid-Connected Photovoltaic Capacity Growth, 1999–2009," Federal Energy Regulatory Commission (June 7, 2010), accessed June 7, 2011, http://www.ferc.gov/market-oversight/othr-mkts/renew/othr-rnw-us-pv-cap-grwth.pdf; U.S. Department of Energy, "Annual Report on U.S. Wind Power Installation, Cost, and Performance Trends: 2006," accessed June 7, 2011, http://www.nrel.gov/docs/fy07osti/41435.pdf.

14. L. D. Frank, S. Kavage, and B. Appleyard, "The Urban Form and Climate Change Gamble," *Planning* 73, no. 8 (2007): 18–23.

15. S. J. Anders, D. O. DeHaan, N. Silva-Send, S. Tanaka, and L. Tyner, "Applying California's AB32 Targets to the Regional Level: A Study of San Diego County," *Energy Policy* 37, no. 7 (2009): 2831–35.

16. R. W. Willson and K. D. Brown, "Carbon Neutrality at the Local Level: Achievable Goal or Fantasy?" *Journal of the American Planning Association* 74, no. 4 (2008): 497–504.

17. European Commission, "The EU Climate and Energy Package," accessed November 2, 2010, http://ec.europa.eu/environment/climat/climate _action.htm.

18. United Nations Framework Convention on Climate Change, "Copenhagen Accord: Draft Decision CP.15" (December 19, 2009), accessed November 2, 2010, https://docs.google.com/viewer?url=http://unfccc.int /resource/docs/2009/cop15/eng/l07.pdf.

19. Johan Rockstrom, Will Steffen, Kevin Noone, Asa Persson, F. Stuart Chapin, Eric F. Lambin, Timothy M. Lenton, et al., "A Safe Operating Space for Humanity," *Nature* 461, no. 7263 (2009): 472–75.

20. CO2Now.org, accessed July 31, 2010, http://www.co2now .org/.

21. Malte Meinshausen, Nicolai Meinshausen, William Hare, Sarah C. B. Raper, Katja Frieler, Reto Knutti, David J. Frame, and Myles R. Allen, "Greenhouse-Gas Emission Targets for Limiting Global Warming to 2°C," *Nature* 458, no. 7242 (April 30, 2009): 1158–62.

22. Intergovernmental Panel on Climate Change, *Climate Change 2007: Impacts, Adaptation and Vulnerability. Contribution of Working Group II to the Fourth Assessment Report of the Intergovernmental Panel on Climate Change*, ed. M. L. Parry, O. F. Canziani, J. P. Palutikof, P. J. van der Linden, and C. E. Hanson (Cambridge: Cambridge University Press, 2007), 776.

23. For examples of state and international targets, see http://www .pewclimate.org/what_s_being_done/targets.

24. City of San Francisco, *Climate Action Plan for San Francisco: Local Actions to Reduce Greenhouse Gas Emissions* (2004).

25. Mayor's Greenprint Denver Advisory Council, *City of Denver Climate Action Plan* (2007).

26. Thomas R. Karl, Jerry M. Melillo, and Thomas C. Peterson, eds., *Global Climate Change Impacts in the United States* (New York: Cambridge University Press, 2009).

Chapter 5

1. "Endangerment and Cause or Contribute Findings for Greenhouse Gases under Section 202(a) of the Clean Air Act," U.S. Environ-

mental Protection Agency, accessed November 8, 2010, http://www.epa.gov/climatechange/endangerment.html.

2. City of Worcester, Climate Action Plan (December 2006).

3. California Air Pollution Control Officers Association (CAPCOA), *Quantifying Greenhouse Gas Emissions Measures: A Resource for Local Government to Assess Emission Reductions from Greenhouse Gas Mitigation Measures* (Sacramento, CA: Author, August 2010), 3.

4. Ibid, pp. 33–34.

5. City of West Hollywood, Climate Action Plan: Public Review Draft (June 2010).

6. *Moving Cooler: An Analysis of Transportation Strategies for Reducing Greenhouse Gas Emissions*, Urban Land Institute, accessed February 28, 2011, http://www.movingcooler.info/.

7. Ibid,, p. 17.

8. Ibid., p. 2.

9. For additional information, see the National Complete Streets Coalition at http://www.completestreets.org/.

10. *Moving Cooler: An Analysis of Transportation Strategies for Reducing Greenhouse Gas Emissions*, Urban Land Institute, accessed February 28, 2011, http://www.movingcooler.info/.

11. "Green Building," U.S. Environmental Protection Agency, accessed February 27, 2011, http://www.epa.gov/greenbuilding/.

12. "U.S. Green Building Council," accessed February 27, 2011, from http://www.usgbc.org/.

13. "Build It Green," accessed February 27, 2011, from http://www.builditgreen.org/.

14. "LED Street Lighting Energy Efficiency Program," Bureau of Street Lighting City of Los Angeles Department of Public Works, accessed March 2, 2011, http://www.ci.la.ca.us/bsl/.

15. G. Kats, L. Alevantis, A. Berman, E. Mills, and J. Perlman, *The Costs and Financial Benefits of Green Buildings* (California Sustainable Building Task Force, 2003).

16. "Energy Star," U.S. Environmental Protection Agency, accessed February 27, 2011, http://www.energystar.gov/.

17. "Home Energy Saver," U.S. Department of Energy, accessed February 27, 2011, http://hes.lbl.gov/consumer/.

18. "Cash for Grass Rebate Program," City of Roseville, CA, accessed February 27, 2011, http://www.roseville.ca.us/eu/water_utility/water_conservation/for_home/cash_for_grass/default.asp.

19. Food and Agriculture Organization of the United Nations [FAO], *Soil Carbon Sequestration for Improved Land Management* (Rome: Author, 2001), ftp://ftp.fao.org/agl/agll/docs/wsrr96e.pdf.

20. Lal, R. "Soil Carbon Sequestration Impacts on Global Climate Change and Food Security," *Science* 304 (2004): 1623–27.

21. Ibid.

22. Intergovernmental Panel on Climate Change, "Summary for Policymakers," in *Climate Change 2007: Impacts, Adaptation and Vulnerability. Contribution of Working Group II to the Fourth Assessment Report of the Intergovernmental Panel on Climate Change*, ed. M. L. Parry, O. F. Canziani, J. P. Palutikof, P. J. van der Linden, and C. E. Hanson (Cambridge: Cambridge University Press, 2007), 5; Pete Smith et al., "Policy and Technological Constraints to Implementation of Greenhouse Gas Mitigation Options in Agriculture," *Agriculture, Ecosystems, and Environment*, 118 (2007): 6–18.

23. H. Steinfeld, P. Gerber, T. Wassenaar, V. Castel, M. Rosales, and C. de Haan, *Livestock's Long Shadow Environmental Issues and Options* (Rome: Food and Agriculture Organization of the United Nations, 2006), 26 pp.

24. Ibid.

25. Ibid.

26. "Food and Yard Waste at Your House," City of Seattle Public Utilities, accessed March 3, 2011, http://www.seattle.gov/util/Services /Yard/Yard_Waste_Collection/index.asp.

27. "Residential Composting," City of San Francisco Environment, accessed March 3, 2011, http://www.sfenvironment.org/our_programs /interests.html?ssi=3&ti=6&ii=225.

28. TerraPass, accessed March 1, 2011, http://www.terrapass.com/.

Chapter 6

1. Intergovernmental Panel on Climate Change, "Summary for Policymakers," in *Climate Change 2007: The Physical Science Basis. Contribution of Working Group I to the Fourth Assessment Report of the Intergovernmental Panel on Climate Change*, ed. S. Solomon, D. Qin, M. Manning, Z. Chen, M. Marquis, K. B. Averyt, M. Tignor, and H. L. Miller (Cambridge and New York: Cambridge University Press, 2007).

2. B. L. Turner, R. E. Kasperson, P. A. Matson, J. J. McCarthy, R. W. Corell, L. Christenson, N. Eckley, J. X. Kasperson, A. Luer, M. L. Martello, C. Polsky, A. Pulsipher, and L. Schiller, "A Framework for Vulnerability Analysis in Sustainability Science," *Proceedings of the National Academy of Sciences of the United States of America* 100, no. 14 (2003): 8074–79.

3. J. Howard, "Climate Change Mitigation and Adaptation in Developed Nations: A Critical Perspective on the Adaptation Turn in Urban Climate Planning," in *Planning for Climate Change,* ed. S. Davoudi, J. Crawford, and A. Mehmood (London: Earthscan, 2009).

4. United Nations Framework Convention on Climate Change, Copenhagen Accord (2009), accessed July 1, 2010, http://unfccc.int/resource /docs/2009/cop15/eng/11a01.pdf#page=4.

5. "Multi-hazard Mitigation Planning," Federal Emergency Management Agency, accessed November 12, 2010, http://www.fema.gov/plan/mitplanning/index.shtm.

6. Marcus Moench, "Adapting to Climate Change and the Risks Associated with Other Natural Hazards: Methods for Moving from Concepts to Action," in *Adaptation to Climate Change*, ed. E. Lisa F. Schipper and Ian Burton (London: Earthscan, 2009), 256.

7. California Natural Resources Agency, *California Climate Adaptation Strategy* (2009), http://www.climatechange.ca.gov/adaptation/.

8. "Safety Tips—Wildfire," San Diego Fire-Rescue Department, accessed November 10, 2010, http://www.sandiego.gov/fireandems/safety/wildfire.shtml.

9. California Natural Resources Agency, *California Climate Adaptation Strategy* (2009), http://www.climatechange.ca.gov/adaptation/.

10. Cal-Adapt, California Energy Commission, Public Interest Energy Research program (2011), http://cal-adapt.org/.

11. Federal Emergency Management Agency (FEMA), *Understanding Your Risks: Identifying Hazards and Estimating Losses,* FEMA 386-2 (Washington, DC: FEMA, 2001).

12. Thomas R. Karl, Jerry M. Melillo, and Thomas C. Peterson, eds., *Global Climate Change Impacts in the United States* (New York: Cambridge University Press, 2009).

13. California Natural Resources Agency, *California Climate Adaptation Strategy* (2009), http://www.climatechange.ca.gov/adaptation/.

14. "Population," National Atmospheric and Oceanic Administration (2009), http://oceanservice.noaa.gov/facts/population.html.

15. S. L. Loiko, "Moscow Death Toll Soars as Heat Wave Persists," *LA Times,* August 10, 2010, http://articles.latimes.com/2010/aug/10/world/la-fg-russia-heat-deaths-20100810.

16. "France Heat Wave Death Toll Set at 14,802," *USA Today,* September 9, 2003, http://www.usatoday.com/weather/news/2003-09-25-france-heat_x.htm.

17. J. Pomfret, "130 Deaths Blamed on California Heat Wave," *Washington Post,* July 29, 2006, http://www.washingtonpost.com/wp-dyn/content/article/2006/07/28/AR2006072801648.html.

18. "Ground-Level Ozone," U.S. Environmental Protection Agency (EPA), accessed November 10, 2010, http://www.epa.gov/air/ozonepollution/health.html.

19. R. de Loe, R. Kreutzwiser, and L. Moraru, "Adaptation Option for the Near Term: Climate Change and the Canadian Water Sector," *Global Environmental Change* 11 (2001): 231–45; B. Smit, I. Burton, R. J. T. Klein, and J. Wandel, "An Anatomy of Adaptation to Climate Change and Variability," *Climatic Change* 45 (2000): 223–51; B. Smit and J. Wandel, "Adaptation,

Adaptive Capacity and Vulnerability," *Global Environmental Change* 16 (2006): 282–92; J. B. Smith, "Setting Priorities for Adapting to Climate Change," *Global Environmental Change* 7, no. 1 (2007): 251–64.

20. List compiled from B. Smit, I. Burton, R. J. T. Klein, and J. Wandel, "An Anatomy of Adaptation to Climate Change and Variability," *Climatic Change* 45 (2000): 223–51; J. B. Smith, J. M. Vogel, and J. E. Cromwell III, "An Architecture for Government Action on Adaptation to Climate Change: An Editorial Comment," *Climate Change* 95 (2009): 53–61.

Chapter 8

1. Sources for this case include the planning documents and an interview with Michael Armstrong, senior sustainability manager, City of Portland Office of Sustainability.

2. "Portland's Green Dividend," CEO for Cities, accessed November 8, 2010, http://www.ceosforcities.org/files/PGD%20FINAL.pdf.

3. City of Portland Bureau of Planning and Sustainability and Multnomah County Sustainability Program, City of Portland and Multnomah County Climate Action Plan (2009), p. 39, accessed November 8, 2010, http://www.portlandonline.com/bps/index.cfm?c=49989&a=268612.

4. Sources for the case included the planning documents and interviews with the mayor of Evanston, Elizabeth Tisdahl; Paige K. Finnegan, chief operating officer at e-One, LLC, and co-chair of the Evanston Environment Board; and Dr. Stephen A. Perkins, senior vice president, Center for Neighborhood Technology.

5. Mayor Tisdahl was elected in 2009. Prior to that she served on the city council when the plan was adopted.

6. City of Evanston, Climate Action Plan (November 2008), 3.

7. Ibid., 6.

8. Ibid., 9.

9. Sources for the case include the planning documents and an interview with Lindsay Baxter, City of Pittsburgh sustainability coordinator.

10. Pittsburgh Climate Initiative, Pittsburgh Climate Action Plan (June 2008), viii.

11. "Pittsburgh Climate Initiative," Green Building Alliance, accessed September 9, 2010, http://www.gbapgh.org/content.aspx?ContentID=93.

12. Sources for this case include the planning documents, an interview with Deborah Nelson, planning manager, City of San Carlos, and press releases from the City of San Carlos: http://www.cityofsancarlos.org/civica/press/display.asp?layout=1&Entry=533; http://www.cityofsancarlos.org/civica/filebank/blobdload.asp?BlobID=6035.

13. "San Carlos Wins Outstanding Planning Award," City of San Carlos Planning Department Press Release (May 11, 2010), http://www.cityofsancarlos.org/civica/press/display.asp?layout=1&Entry=533.

14. "San Carlos General Plan 2030 Wins Visionary Award," City of San Carlos Planning Department Press Release (March 31, 2010), http://www.cityofsancarlos.org/civica/press/display.asp?layout=1&Entry=525.

15. "San Carlos General Plan 2030 Wins Visionary Award," City of San Carlos Planning Department Press Release (March 31, 2010), http://www.cityofsancarlos.org/civica/press/display.asp?layout=1&Entry=525.

16. Sources for this case include the planning documents and interviews with Susanne M. Torriente, sustainability director (plan leader), and Amy Knowles, organizational development administrator, Department of Environmental Resources Management (DERM) (plan coordinator).

17. Miami-Dade County, A Long Term CO_2 Reduction Plan for Miami-Dade County, Florida, (December 2006), accessed June 1, 2011, http://www.miamidade.gov/derm/library/air_quality/CO2_Reduction_Final_Report.pdf.

18. "Comprehensive Development Master Plan (CDMP)," Miami-Dade County, accessed June 1, 2011, http://www.miamidade.gov/planzone/planning_metro_CDMP.asp.

19. Sources for this case include the planning documents and an interview with Anne Marie Holen, special projects coordinator, City of Homer.

20. City of Homer, City of Homer Climate Action Plan (December 2007), accessed February 23, 2011, http://www.cityofhomer-ak.gov/sites/default/files/fileattachments/climate_action_plan_0.pdf.

21. City of Homer, City of Homer Climate Action Plan Implementation Project Final Report (December 2009), accessed February 23, 2011, http://www.sustainablehomer.org/climate_change_files/CAP%20Implementation%20Final%20Report%2012-14-09.pdf

22. "Sustainability," City of Homer, accessed March 2, 2011, http://www.cityofhomer-ak.gov/community/sustainability.

23. "Money, Energy, and Sustainability," City of Homer, accessed March 2, 2011, http://www.cityofhomer-ak.gov/sites/default/files/fileattachments/employee_sustainibility_policy_guide.pdf.

24. City of Homer, Comprehensive Plan 2008 (adopted 2010), accessed March 3, 2011, http://www.cityofhomer-ak.gov/planning/comprehensive-plan-2008-adopted-2010.

25. City of Homer, City of Homer 2011–2016 Capital Improvement Plan (November 2010), accessed March 3, 2011, http://www.cityofhomer-ak.gov/sites/default/files/fileattachments/City%20of%20Homer%202011-2016%20CIP.pdf.

26. City of Homer, Homer Comprehensive Economic Development Strategy (December 2011),, accessed March 4, 2011, http://www.city ofhomer-ak.gov/sites/default/files/fileattachments/ceds_feb_2011_final.pdf.

Appendix A

1. "Organization," Intergovernmental Panel on Climate Change, accessed May 10, 2010, http://www.ipcc.ch/organization/organization.htm.

2. "Program Overview," U.S. Global Change Research Program, accessed September 13, 2010, http://www.globalchange.gov/about.

3. The agencies and departments are Agency for International Development; National Oceanic and Atmospheric Administration; National Institutes of Health; United States Geological Survey; Environmental Protection Agency; National Aeronautics and Space Administration; National Science Foundation; Smithsonian Institution; Department of Agriculture; Department of Defense; Department of Energy; Department of State; Department of Transportation; and Department of the Interior.

4. Intergovernmental Panel on Climate Change, "Summary for Policymakers," in *Climate Change 2007: The Physical Science Basis. Contribution of Working Group I to the Fourth Assessment Report of the Intergovernmental Panel on Climate Change*, ed. S. Solomon, D. Qin, M. Manning, Z. Chen, M. Marquis, K. B. Averyt, M. Tignor, and H. L. Miller (Cambridge and New York: Cambridge University Press, 2007), 5.

5. Ibid.

6. D. S. Arndt, M. O. Baringer, and M. R. Johnson, eds., "State of the Climate in 2009," *Bulletin of the American Meteorological Society* 91, no. 6 (2010): S1–S224.

7. Ibid., S26.

8. Robert Henson, *The Rough Guide to Climate Change*, 2nd ed. (New York, NY: Rough Guides, 2008), 6.

9. U.S. Environmental Protection Agency, *Inventory of U.S. Greenhouse Gas Emissions and Sinks: 1990–2008*, EPA 430-R-10-006 (Washington, DC: Author, 2010).

10. S. Solomon, D. Qin, M. Manning, R. B. Alley, T. Berntsen, N. L. Bindoff, Z. Chen, A. Chidthaisong, J. M. Gregory, G. C. Hegerl, M. Heimann, B. Hewitson, B. J. Hoskins, F. Joos, J. Jouzel, V. Kattsov, U. Lohmann, T. Matsuno, M. Molina, N. Nicholls, J. Overpeck, G. Raga, V. Ramaswamy, J. Ren, M. Rusticucci, R. Somerville, T. F. Stocker, P. Whetton, R. A. Wood, and D. Wratt, "Technical Summary," in *Climate Change 2007: The Physical Science Basis. Contribution of Working Group I to the Fourth Assessment Report of the Intergovernmental Panel on Climate Change*, ed. S. Solomon, D. Qin, M. Manning,

Z. Chen, M. Marquis, K. B. Averyt, M. Tignor and H. L. Miller (Cambridge and New York: Cambridge University Press, 2007).

11. Committee on Stabilization Targets for Atmospheric Greenhouse Gas Concentrations, *Climate Stabilization Targets: Emissions, Concentrations, and Impacts over Decades to Millennia* (Report in Brief) (Washington, DC: National Academies Press, 2010).

12. U.S. Environmental Protection Agency, *Inventory of U.S. Greenhouse Gas Emissions and Sinks: 1990–2008*, EPA 430-R-10-006 (Washington, DC: Author, 2010).

13. Summary of impacts and consequences is based on the following reports: Thomas R. Karl, Jerry M. Melillo, and Thomas C. Peterson, eds., *Global Climate Change Impacts in the United States* (New York: Cambridge University Press, 2009); National Research Council, *Adapting to the Impacts of Climate Change* (Washington DC: National Academies Press, 2010).

14. Compiled from the U.S. EPA (http://www.epa.gov/climate change/glossary.html) and California Energy Commission (http://www .climatechange.ca.gov/glossary/index.html).

15. V. Ramaswamy, O. Boucher, J. Haigh, D. Hauglustaine, J. Haywood, G. Myhre, T. Nakajima, G. Y. Shi, S. Solomon, "Radiative Forcing of Climate Change" (http://www.grida.no/publications/other/ipcc_tar/), in *Climate Change 2001: The Scientific Basis. Contribution of Working Group I to the Third Assessment Report of the Intergovernmental Panel on Climate Change*, ed. J. T. Houghton, Y. Ding, D. J. Griggs, M. Noguer, P. J. van der Linden, X. Dai, K. Maskell, and C. A. Johnson (Cambridge and New York: Cambridge University Press), 881.

Index